Field Manual
No. 7-100

Headquarters
Department of the Army
Washington, DC,1 May 2003

# OPFOR Opposing Force Doctrinal Framework and Strategy

## Contents

# FOREWORD

In today's complicated and uncertain world, it is impossible to predict the exact nature of future conflict that might involve the U.S. Army. So the Army must be ready to meet the challenges of any type of conflict, in all kinds of places, and against all kinds of threats. This is the nature of the contemporary operational environment (COE), and training for such an environment requires a different type of Opposing Force (OPFOR) than that of the past.

The Deputy Chief of Staff for Intelligence (DCSINT) of the U.S. Army Training and Doctrine Command (TRADOC) is the Executive Agent for the development, management, administration, integration, and approval functions of the OPFOR Program across the Army. Thus, the TRADOC DCSINT is responsible for documenting the doctrine, organization, and capabilities of a contemporary OPFOR that is appropriate for training the Army's leaders, soldiers, and units for the COE.

In the FM 7-100 series, the TRADOC Office of the Deputy Chief of Staff for Intelligence (ODCSINT) has created a flexible baseline for an OPFOR that can be adapted to meet a variety of different training requirements in a number of different scenarios that reflect the COE. The OPFOR doctrinal framework and strategy outlined in FM 7-100 represent a realistic composite of potential adversaries the Army might encounter in the real-world situations of the foreseeable future. However, the world is continually changing, as are the threats and challenges for which the Army must be prepared. The Army must remain flexible, as must the OPFOR designed to serve as a challenging sparring partner in the training environment.

This manual is approved for use in all Army training venues. However, as the contemporary OPFOR and other aspects of the COE are integrated into Army training, the TRADOC ODCSINT and the intelligence community will continue research and analysis of real-world developments and trends. The goal of this continued effort is to keep our OPFOR and our understanding of the COE truly contemporary and relevant as the world around us changes. Thus, this manual is intended to be a living document, and the ODCSINT will modify and change it as often as necessary in order to ensure its continued relevance in light of changes and developments in the COE. In anticipation of such changes, this manual will be published primarily in electronic format with only limited distribution of hard-copy, printed manuals. Users need to monitor the Reimer Digital Library (http://www.adtdl.army.mil) and the DCSINT-Threats Web site (http://www.leavenworth.army.mil/threats/index) for periodic updates.

Maxie L. McFarland

**MAXIE L. MCFARLAND**
Deputy Chief of Staff for Intelligence
U.S. Army Training and Doctrine Command

# Preface

This manual is one of a series that describes a contemporary Opposing Force (OPFOR) for training U.S. Army commanders, staffs, and units. See the Reference section for a list of the manuals in this series. Together, these manuals outline an OPFOR than can cover the entire spectrum of military and paramilitary capabilities against which the Army must train to ensure success in any future conflict.

Applications for this series of manuals include field training, training simulations, and classroom instruction throughout the Army. All Army training venues should use an OPFOR based on these manuals, except when mission rehearsal or contingency training requires maximum fidelity to a specific country-based threat. Even in the latter case, trainers should use appropriate parts of the OPFOR manuals to fill information gaps in a manner consistent with what they do know about a specific threat.

The proponent for this publication is HQ TRADOC. Send comments and recommendations on DA Form 2028 directly to the OPFOR and Threat Integration Directorate of the TRADOC Office of Deputy Chief of Staff for Intelligence at the following address: Director, OPFOR and Threat Integration Directorate, ATTN: ATIN-T (Bldg 53), 700 Scott Avenue, Fort Leavenworth, KS 66027-1323.

This publication is available on the Reimer Digital Library at http://www.adtdl.army.mil. Readers should monitor that site and also the TRADOC DCSINT-Threats World Wide Web (WWW) site at http://www.leavenworth.army.mil/threats/index for the status of this manual and information regarding updates. Periodic updates, subject to the normal approval process, will occur as a result of the normal production cycle in accordance with TRADOC regulation 25-36, paragraphs 2-17 and 4-7. The date on the cover and title page of the electronic version will reflect the latest update.

Unless this publication states otherwise, masculine nouns or pronouns do not refer exclusively to men.

### Economics

The economic variable establishes the boundaries between the "haves" and the "have-nots." This gap of economic differences among nation-states and other actors can cause conflict. Economic superiority, rather than military superiority, may be the key to power or dominance within a region. However, economic position often represents a nation or non-state actor's ability to buy military technology or to conduct prolonged operations.

Economics help define the relationship between a nation or non-state actor and other actors at the regional or global level. These regional or global economic relationships could result in military or political assistance.

### Sociological Demographics

The demographics variable includes the cultural, religious, and ethnic makeup of a given region, nation, or non-state actor. Extreme devotion to a particular cause or significant hatred of a particular group may provide an enemy with an unshakable will and a willingness to die for the cause. U.S. forces may also find that large segments of the population around them are sympathetic to the same cause as the enemy force. The needs of the local population can create heavy demands on U.S. military units, particularly their supply and medical systems. Refugees and internally displaced persons may increase the complexity of the environment. The enemy may use civilians as shields or obstacles or as cover for hostile intelligence services.

### Information

Media and other information means can make combat operations transparent to the world, visible to all who have access to data. Various actors seek to use perception management to control and manipulate how the public sees things. They will exploit U.S. mistakes and failures and use propaganda to sway the local population to support their cause. Media coverage can impact on U.S. political decision making, international opinion, or the sensitivities of coalition members.

Even without sophisticated sensors and information systems, actors native to the area or region often have greater situational awareness than U.S. forces. Various actors are able to access commercial systems (such as satellite communications and imagery) for the larger picture. For a more detailed view, they can use human networks operating over normal telephone lines or with cellular telephones to maintain situational awareness.

### Physical Environment

The main elements in the physical environment are terrain and weather. Potential enemies clearly understand that less complex and open environments favor a U.S. force with its long-range, precision-guided weapons and sophisticated reconnaissance capability. So they will try to avoid the types of operations and environments for which such U.S. forces are optimized. They will try to operate in urban areas and other complex terrain and in weather conditions that may adversely affect U.S. military operations and mitigate technological advantages.

### Technology

The technology that nations or non-state actors can bring to the OE includes what they can develop and produce, as well as what they could import. Access to technological advances available on the global market is slowly eating away at the technological advantage the United States has enjoyed in the past.

It is likely that some high-end forces in a particular region of the world could field a few systems that are more advanced than those of the U.S. force deployed there. Easy access to new technology allows potential adversaries to achieve equality or even overmatch U.S. systems in selected niche areas. Many countries are trying to acquire relatively low-cost, high-payoff, new technologies. In addition, upgrades and hybridization allow older systems to compete with more modern capabilities, thus neutralizing the technical advantage of many modern forces. In urban areas or other complex terrain, less advanced systems may still find effective uses. Various actors may find adaptive and innovative ways of using systems for other than their originally intended applications.

### External Organizations

When the U.S. Army goes into a failed state or into areas torn by conflict, it is likely to find international humanitarian relief organizations at work there. These external organizations continue to grow in influence and power, as well as in willingness to become involved in crisis situations that were previously purely military operations. These external organizations can have both stated and hidden interests and objectives that can either assist or hinder U.S. mission accomplishment. The presence of transnational corporations operating in a country or region can also place added pressure on U.S. forces to avoid collateral damage to civilian life and property. U.S. forces may have to divert troops and resources from their assigned missions to conduct rescues or provide security for various external organizations.

### National Will

The variable of national will reflects how much each country's people and government are behind what the military or paramilitary forces are doing. This can influence the objectives of a conflict, its duration, and the conditions for ending it.

A country will try to attack its opponent's national will and still preserve its own. Clearly, most foreign countries view U.S. national will as a point of vulnerability. Thus, a potential adversary may perceive the collective will of his people as a comparative advantage against the United States.

History has proven that battlefield victory does not always go to the best-trained, best-equipped, and most technologically advanced force. Victory often goes to the side that most wants to win, needs to win, and is willing to sacrifice to do so.

### Time

In most cases, potential opponents of the United States view time as being in their advantage. When U.S. forces have to deploy into the area over long time and distance, the opponent can use this time to adjust the nature of the conflict to something for which the U.S. forces are not prepared.

First, the opponent will try to control the entry of U.S. forces into the area. If access control fails, the enemy still has the opportunity to oppose lightly equipped

U.S. early-entry units and try to prevent full deployment of the rest of the force. The opponent will try to speed up the tempo, to rapidly defeat its local or regional enemy or to defeat U.S. early-entry forces before the United States can deploy overwhelming military power. If that fails, the opponent will try to prolong the conflict and to outlast the U.S. will to continue.

### Military Capabilities

Military capabilities of a nation-state or non-state actor are measured in relative terms, in comparison to the capabilities of other actors against which they might be applied. Most of the military forces in the world continue to operate in conventional ways, which remain sufficient against other local and regional actors.

However, once the United States becomes involved, these same military forces may have to use adaptive or asymmetric approaches. Various nations and other foreign entities around the world study the United States and its military forces. They generally view the United States as a major power—the world's only superpower—with an overall advantage in technology and warfighting capability. Despite these strengths, other actors see some weaknesses that they may be able to exploit. They can use these perceptions as a guide to optimizing the effectiveness of their own forces and to find ways to negate current U.S. advantages.

Military capabilities may be the most critical and the most complex variable that affects military operations. However, the military variable does not exist in isolation from the other variables that help determine the overall OE. It interacts with the other variables, and all the other variables can affect military capabilities. Potential enemies can use any or all of these factors against the Army as it tries to accomplish its missions in various parts of the world or in various training environments.

## REAL WORLD

In the real world, the COE is the entire set of conditions, circumstances, and influences that U.S. Armed Forces can expect to face when conducting military operations to further the national interests of the United States, its friends, and allies. The COE is "contemporary" in the sense that it does not represent conditions that existed only in the past or that might exist only in the remote future, but rather those conditions that exist today and in the clearly foreseeable, near future. This COE consists not only of the military and/or paramilitary capabilities of potential real-world adversaries, but also of the manifestations of the ten other variables that help define any OE.

## TRAINING

In training environments, the COE is the OE created to approximate the demands of the real-world COE and to set the conditions for desired training outcomes. This involves the appropriate combination of an OPFOR (with military and/or paramilitary capabilities representing a composite of a number of potential adversaries) and other OE variables in a realistic, feasible, and plausible manner. The purpose of the COE in training simulations is to produce the necessary training outcomes.[1]

[1] The same type of COE conditions can be created to support some combat development activities that do not require simulation of a specific real-world potential adversary. However, some combat development activities may require portrayal of an OE that extends further into the future than is typical for the COE.

Even in the COE for training, it is possible to speak of an overall COE that addresses the qualities of virtually any OE in which the units or individuals being trained might be called upon to operate. In this sense, there are the same "constants" as in the real-world COE.

### INTERACTION AND LINKAGE OF VARIABLES

The variables of the COE do not exist in isolation from one another. The linkages of the variables cause the complex and often simultaneous dilemmas that a military force might face. In order to provide realistic training, training scenarios must try to simulate this synergistic effect to the maximum degree that is feasible.

The COE is not just about the OPFOR. The COE variables and their interaction provide the robust environment and context for OPFOR operations. The complexity of the specific OE in training can be adjusted to keep it appropriate for the required training objectives and the training state of various U.S. Army units.

### ADAPTIVE AND CHANGING

The nature of the COE is adaptive and constantly changing. As the United States and its military forces interact with the COE in a real-world sense, the OE changes. As the Army applies the lessons learned from training in a COE setting, the OPFOR and potential real-world adversaries will also learn and adapt.

The development of the COE for training started with research to develop an understanding of the real-world COE and trends that affect military operations. Then, taking into consideration the desired training outcomes and leader development goals, the authors of the FM 7-100 series proceeded to document an OPFOR doctrine and structure that reflect the real-world COE, and the Army began integrating this OPFOR and other COE variables into training scenarios. Meanwhile, the authors of the FM 7-100 series are continuing to research the real-world COE and to mature the OPFOR and the COE in training in order to provide a richer, appropriately challenging training environment and keep the OPFOR and the COE truly "contemporary."

## ENEMY, THREAT, AND OPFOR

Before going further into the COE, the contemporary OPFOR, and the intended uses of this manual, it may be useful to define some key terms and the distinctions among them. It is important to distinguish among the terms *enemy*, *threat*, and *OPFOR* and to use them correctly.

### ENEMY

The U.S. Army defines *enemy* as "the individual, group of individuals (organized or not organized), paramilitary or military force, national entity, or national alliance that is in opposition to the United States, its allies, or multinational partners." In other words, the *enemy* is whoever is actually opposing the United States in a particular conflict.[2] Thus, this term is synonymous with adversary or opponent.

---

[2] This definition of *enemy* is from the U.S. point of view. After this Introduction, the chapters of this manual address their topics from the OPFOR point of view. So, *friendly* refers to the OPFOR and its allies, and enemy refers to the enemy of the OPFOR, which may be an opponent within its own country or region or an extraregional opponent (normally the United States or a U.S.-led coalition).

## NATION-STATE ACTORS

Nation-states fall into four basic categories according to their roles in the international community. The categories are core states, transition states, rogue states, and failed or failing states.

The category of *core states* includes more than half of the nearly 200 countries in the world today. These are basically democratic (although to varying degrees) and share common values and interests. Within this larger group, there is an "inner core" of major powers. These are the advanced countries, including the United States, that generally dominate world politics. Most conflict with global consequences will involve the core states in some fashion or another.

*Transition states* are other larger, industrial-based countries—mostly emerging regional powers—that are striving to become major powers. High-end transition states are moving from an industrial-based society to an information-based society. Low-end transition states are seeking to move from an agricultural-based society to an industrial base. As states try to make this transition, there are cycles of political stability and instability, and the outcome of the transition is uncertain. Some transition states may successfully join the ranks of core states and even become major powers within that context; others may become competitors.

*Rogue states* are those that are hostile to their neighbors or to core states' interests. These countries can sponsor international terrorism or even confront U.S. military forces operating in the region. *Failed or failing states* are fragmented in such a way that a rule of law is absent; their instability is a threat to their neighbors and the core states.

Countries can move from one category to another, as conditions change. Sometimes countries join together in multinational alliances and coalitions. Together, they have more strength and can become a power to be reckoned with.

## NON-STATE ACTORS

Non-state actors are those that do not represent the forces of a particular nation-state. Such non-state elements include rogue actors as well as third-party actors.

Like rogue states, *rogue actors* are hostile to other actors; however, they may be present in one country or extend across several countries. Examples include insurgents, guerrillas, mercenaries, and transnational or subnational political movements. Particular sources of danger are terrorists and drug-trafficking or criminal organizations, since they may have the best technology, equipment, and weapons available, simply because they have the money to buy them. These non-state rogue actors may use terror tactics and militarily unconventional methods to achieve their goals.

*Third-party actors* may not be hostile to other actors. However, their presence, activities, and interests can affect the ability of military forces to accomplish their mission when operating in a foreign country. These third-party actors can be refugees, internally displaced persons, and other civilians on the battlefield, including international humanitarian relief agencies, transnational corporations, and the news media. These individuals and groups bring multiple sources of motivation, ideology, interests, beliefs, or political affiliations into consideration. They may be sources of civil unrest. Their presence may require military forces to consider the potential impacts of traffic congestion, demonstrations, sabotage, and information manipulation.

### REAL-WORLD AND TRAINING CONSIDERATIONS

When U.S. forces become involved in a particular country or region, they must take into account the presence and influence of these various types of threats and other actors. In a training environment, an OPFOR can represent a composite of those nation-state or non-state actors that constitute military and/or paramilitary forces that could present a threat to the United States, its friends, or its allies. Other, non-state actors that fall in the category of nonmilitary forces or elements are not part of the OPFOR, but could be part of the COE used in the training environment.

## CONTEMPORARY OPFOR

This manual describes the doctrinal framework and strategy of a flexible, thinking, adaptive, contemporary OPFOR that applies its doctrine with considerable initiative. (See the definition of *contemporary OPFOR* above.) It is applicable to the entire training community, including the OPFORs at all of the combat training centers (CTCs), the TRADOC schools, and units in the field. It provides an OPFOR that believes that, through adaptive use of all available forces and capabilities, it can create opportunities that, properly leveraged, can allow it to fight and win, even against a technologically superior opponent such as the United States.

### BASELINE

As a baseline for developing specific OPFORs for specific training environments, this manual describes an OPFOR that is representative of contemporary nation-states. This composite of the characteristics of real-world military and paramilitary forces provides a framework for the realistic and relevant portrayal of capabilities and actions that U.S. armed forces might face in the COE.

For this composite of real-world threats, the manual refers to the country in question as "**the State**."[5] It describes this artificial country in terms of the eleven critical variables of the COE. As the baseline for the contemporary OPFOR that is representative of real-world forces, the State is not a peer competitor of the United States. However, it is a dominant power in its region of the world and is capable of challenging U.S. interests there. The general characteristics of the State could fit a number of different types of potential adversaries in a number of different scenarios.

Like most countries in the world, the State does not design its forces just to fight the United States. It designs them principally to deal with regional threats and to take advantage of regional opportunities. Therefore, the State's national security strategy (including its doctrine, force design, and investment strategy) focuses primarily on maintaining and expanding its position as a regional power. It develops its military forces in a way that ensures conventional power superiority over any of its regional neighbors. These forces, together with the State's other instruments of power, make it a dominant force in its region.

---

[5] In specific U.S. Army training environments, the generic name of the State may give way to other (fictitious) country names such as Atlantis, Upper Flambokia, or Westland.

### Contemporary OPFOR

> **Contemporary OPFOR**
>
> **A plausible, flexible military and/or paramilitary force representing a composite of varying capabilities of actual worldwide forces, used in lieu of a specific threat force, for training and developing U.S. forces.**

Training U.S. forces for the COE requires a different kind of OPFOR from that of the past. The contemporary OPFOR must be less predictable and not based on the armed forces of a particular country. In today's world, the U.S. Army must be prepared to go into any OE and perform its full range of missions. It must be ready to do so in the face of a wide variety of possible threats and at the same time be prepared to deal with third-party actors that may have other interests. Not all threats are purely military in nature. Therefore, the U.S. Army now defines an *OPFOR* as "a plausible, flexible military and/or paramilitary force representing a composite of varying capabilities of actual worldwide forces, used in lieu of a specific threat force, for training and developing U.S. forces."

Thus, in some training environments, a military force alone may be the OPFOR. In other cases, military forces may have paramilitary forces acting in loose affiliation with them, or acting separately from them within the same training environment. These relationships depend on the scenario, which is driven by training requirements.

Various agencies and experts have different lists of real-world threats the United States might have to face. If the U.S. Army were to pick any one of these threats as *the* threat against which to train, that threat would almost certainly not be the one it would actually fight. What is needed is a composite that is representative of the full range and variety of possible threats and OEs. It must have a bit of everything—it could be virtually anybody, anywhere. Therefore, this manual defines this representative composite in a way that is flexible enough to fit the most demanding U.S. Army training requirements and provides a framework for training that creates the leaders, soldiers, and unit skills necessary for success on the next battlefield.

## CONTEMPORARY THREATS AND OTHER ACTORS

There are many types of actors or participants in today's complex world environment. Some of the actors are countries (also called nation-states) and some are not. Nation-states are still dominant actors. However, some power is shifting to nontraditional actors and transnational concerns. There are many potential challenges to traditional concepts like balance of power, sovereignty, national interest, and roles of nation-state and non-state actors.

Of course, not all actors are threats. To be a threat, a nation or organization must have both the capabilities and the *intention* to challenge the United States. The capabilities in question are not necessarily purely military, but encompass all the elements of power available to the nation or organization.

## THREAT

A potential adversary is sometimes designated as a threat. In this sense, the Army defines *threat* as "any specific foreign nation or organization with intentions and military capabilities that suggest it could become an adversary or challenge the national security interests of the United States or its allies." Once hostilities actually begin, the threat becomes the enemy.

## OPPOSING FORCE

An *Opposing Force (OPFOR)* is a training tool that should allow the U.S. Army to train against a challenging and plausible sparring partner that represents the wide range of possible opponents the Army could face in actual conflict. It enables training of all arms of the Army and prepares the Army for potential combat operations.[3]

During the road to war leading up to events in a training scenario, the OPFOR may play the role of a "threat" (potential enemy) that is on the verge of becoming an enemy. However, the actual training event usually deals with a state of hostilities. Thus, once hostilities begin in the training event, the OPFOR acts as the "enemy" of the U.S. force in the training environment.[4]

During the Cold War period, the Army employed OPFORs based on specific real-world threats. However, the Army needs a different type of OPFOR to meet its training requirements for the COE.

### Cold War OPFOR

When the Army established its OPFOR program in 1976 with Army Regulation 350-2, it could hardly have envisioned today's computerized constructive and virtual simulations, or even the evolving requirements of live simulations. It defined an *OPFOR* simply as "an organized force created by and from U.S. Army units to portray a unit of a potential adversary armed force." Thus, all OPFORs were originally threat-based, in the sense that they replicated the forces, capabilities, and doctrine of a particular country officially recognized as a threat or potential adversary. In the midst of the Cold War, the 1976 regulation identified only one potential adversary against which to train: the Soviet Union; by 1983, a revision of the regulation added North Korea as a second threat for replication by an OPFOR. Over time, the Army developed other OPFORs to replicate other threats emerging in places ranging from Latin America and Southwest Asia.

In its time, the threat-based OPFOR served the Army very well, particularly for units targeted against specific threats. The benefits of this training were borne out, for example, in Operation Desert Storm. Techniques and doctrine, including deep attack and the intelligence preparation of the battlefield, developed to cope with specific threats and honed against the OPFOR, enabled the Army to achieve decisive results on the battlefield. However, the OE is dynamic, and the pace of that dynamism has increased with the end of the Cold War and the rapid advancement of information technology.

---

[3] Although the OPFOR is primarily a training tool, it may be used for other purposes. For example, some combat development activities that do not require simulation of a specific real-world potential adversary may use an OPFOR to portray the "threat" or "enemy."

[4] From the OPFOR point of view, its leadership plans and develops forces and methods to deal with one or more threats to its own interests, goals, or survival.

## CRITICAL VARIABLES

Any OE, in the real world or in the training environment, can be defined in terms of eleven critical variables. While these variables can be useful in describing the overall (strategic) environment, they are most useful in defining the nature of specific OEs. Each of these "conditions, circumstances, and influences" and their possible combinations will vary according to the specific situation. In this sense, they are "variables." These variables are interrelated and sometimes overlap. Different variables will be more or less important in different situations. Each OE is different, because the content of the variables is different. Only by studying and understanding these variables—and incorporating them into its training—will the U.S. Army be able to keep adversaries from using them against it or to find ways to use them to its own advantage.

**Critical Variables of COE**

- Nature and Stability of the State.
- Regional and Global Relationships.
- Economics.
- Sociological Demographics.
- Information.
- Physical Environment.
- Technology.
- External Organizations.
- National Will.
- Time.
- Military Capabilities.

### Nature and Stability of the State

It is important to understand the nature and stability of the state (or states) with which or in which the conflict takes place. Study of this variable measures how strong or weak a country is and determines where the real strength of the state lies; it may be in the political leadership, the military, the police, or some other element of the population. Understanding this variable will allow U.S. forces to better understand the nature of the military campaign and the true aims of an enemy campaign, operation, or action. It also helps determine what kinds of threats may be present in a particular country. The real threat to U.S. forces may come from elements other than the military.

### Regional and Global Relationships

Nation-states and/or non-state actors often enter into relationships, which can be regional or global. These partnerships support common objectives, which can be political, economic, military, or cultural. An actor's membership or allegiance to such a relationship can determine its actions of support and motivation. Virtually all conflict will occur with alliances and coalitions, some involving the United States and some involving its adversaries. When actors create regional or global alliances, it can add to their collective capability and broaden the scale of operations and actions.

As the world moves away from the traditional long-term, fixed alliances of the past, regional and global relationships are much more fluid and unpredictable. The choice of a state to be nonaligned does not mean that it will not become involved in a conflict or crisis. It simply means that the state does not make a commitment to another state, alliance, or cause before a situation arises. This lack of precommitment makes it difficult to predict how actors and forces may align when a situation does arise. Alliances can form or change rapidly, even during the course of an operation or campaign.

# Introduction

This manual is the capstone document for the FM 7-100 series, which describes a **contemporary Opposing Force (OPFOR)** that exists for the purpose of training U.S. forces for potential combat operations. This OPFOR reflects the characteristics of military and paramilitary forces that may be present in the **contemporary operational environment (COE)**. Like those real-world threats, the OPFOR will continue to present new and different challenges for U.S. forces. The COE is **constantly changing**, and it is important for U.S. Army training environments to keep pace with real-world developments.

## CONTEMPORARY OPERATIONAL ENVIRONMENT

> **Contemporary Operational Environment (COE)**
>
> **The operational environment that exists today and for the clearly foreseeable future.**

The DOD officially defines an *operational environment* (OE) as "a composite of the conditions, circumstances, and influences that affect the employment of military forces and bear on the decisions of the unit commander" (JP 1-02). The *contemporary operational environment* (COE) is the operational environment that exists today and for the clearly foreseeable future. There are some "constants" or common threads that define the general nature of this COE:

- The United States in not likely to have a peer competitor until 2020 or beyond.
- However, nations will continue to field armed forces and use these forces as a tool to pursue national interests.
- As nations use their armed forces (or other instruments of national power) in pursuit of national interests, their actions may cause U.S. intervention, either unilaterally or as a coalition partner, with or without United Nations mandate.
- Nations that believe the United States may act to counter their national interests will develop diplomatic, informational, economic, and military plans for managing U.S. intervention.
- Nations will continue to modernize their armed forces within the constraints of their economies, but in ways that may negate U.S. overmatch.
- Advanced technology will be available on the world market for a wide variety of nation-state and non-state actors.
- Non-state actors will play an important role in any regional conflict—as combatants or noncombatants.
- All combat operations will be significantly affected by a number of variables in the environment beyond simple military forces.

Thus, one of the constants is that there are variables. Those "variables" in the COE result in a number of different OEs that can occur in specific circumstances or scenarios.

At the same time, the State is aware that aggressive pursuit of its regional goals might lead to intervention by a major power, such as the United States, from outside the region. To the extent possible, therefore, it invests in technologies and capabilities that have utility against both regional and extraregional opponents. The basic force structure of the OPFOR is the same for either type of threat. The State must go to war—or continue the war after extraregional intervention—with whatever it had going into the war.

When an extraregional power intervenes with sufficient force to overmatch the State's, the State has to adapt its patterns of operation. It realizes that the forces and technology that allow it to dominate its neighbors may not be a match for the modern, high-technology forces of a wealthy extraregional power like the United States—at least not in a head-to-head conventional confrontation. However, it can use those means in creative and adaptive ways. To the maximum extent possible, the State plans and trains for adaptive operations and how it will make the transition to them. It is the combination of the State's capabilities and its adaptive strategy, operations, and tactics that make it believe it can take on such an extraregional force and win.

At the strategic level, the State's ability to challenge U.S. interests includes not only the military and paramilitary forces of the State, but also the State's diplomatic-political, informational, and economic instruments of power. Rarely would any country engage the United States or a U.S.-led coalition with purely military means. It is also possible that the State could be part of an alliance or coalition, in which case the OPFOR could include allied forces. These nation-state forces may also operate in conjunction with non-state actors such as insurgents, terrorists, and drug or criminal organizations.

The FM 7-100 series, as a whole, covers not only the military and paramilitary forces of the State, but also other, non-state paramilitary and nonmilitary organizations present in the State's region of the world. An extraregional power becoming involved in that region may have to deal with any or all of these types of military, paramilitary, and nonmilitary elements. It might encounter these elements individually or, more likely, in combination with other such elements. Whether these elements operate in concert or independently, they are an important part of the COE.

Trainers need to consider the total OE and all instruments of power at the disposal of the State and the OPFOR—not just the military element, but also diplomatic-political, informational, and economic means. For a nation-state, these are instruments of national power. For non-state actors whose forces are paramilitary in nature, the other three instruments of power are generally present to one degree or another. Together, these instruments represent the power that actors can bring to bear against the United States.

## FLEXIBILITY

As a training tool, the OPFOR must be a challenging, uncooperative sparring partner, capable of stressing any or all battlefield operating systems of the U.S. force. However, it also must be tailored to meet training requirements.

In the OPFOR baseline presented in this manual, the FM authors often say that the State or the OPFOR "may" be able to do something or "might" or "could" do something. They often use the progressive forms of verbs to say that the State has a "growing" economy or "is developing" a capability or "is continually modernizing." The State participates in the global market, which can allow it to acquire things it cannot produce domestically. Such descriptions give scenario writers considerable flexibility in determining what the State or the OPFOR actually has at a given point in time or a given place on the battlefield—in a particular scenario.

The composite example of this baseline may meet the OPFOR requirements for many U.S. Army training environments. For cases that require an OPFOR based on a type of nation-state with characteristics different from those of the State described in this manual, this baseline provides a framework from which trainers can develop an OPFOR appropriate for their particular training requirements.

The OPFOR must be flexible enough to fit various training requirements. It must be scalable and tunable. Depending on the training requirement, the OPFOR may be a large, medium, or small force. Its technology may be state-of-the-art, relatively modern, obsolescent, obsolete, or an uneven combination of those categories. Its ability to sustain operations may be limited or robust.

## THINKING

This manual describes how the OPFOR thinks, especially how it thinks about fighting its regional neighbors and/or the United States. This thinking determines basic OPFOR strategy—as well as operations and tactics, which are the subjects of other manuals in this series. It drives OPFOR organizational structures and equipment acquisition or adaptation. It also determines how the nation-state OPFOR that represents the armed forces of the State would interact with other, non-state actors that may be present in the COE.

Just because the U.S. force knows something about how the OPFOR has fought in the past does not mean that the OPFOR will always continue to fight that way. A thinking OPFOR will learn from its own successes and failures, as well as those of its potential enemies. It will adapt its thinking, its makeup, and its way of fighting to accommodate these lessons learned. It will continuously look for innovative ways to deal with the United States and its armed forces.

## ADAPTABILITY

Like all military forces, the OPFOR has a basic, conventional design for dealing with forces with capabilities equal to or inferior to its own. Prior to a U.S. force becoming involved, therefore, the OPFOR can use the application or threat of application of that conventional design to dominate or influence its regional neighbors. The OPFOR plans these operations well in advance and tries to execute them as rapidly as possible, in order to preclude regional alliances or outside intervention.

The OPFOR has developed its doctrine, force structure, and capabilities with an eye toward employing them against both regional and extraregional opponents, if necessary. It has thought about and trained for how to adapt once an extraregional force becomes engaged. It has included this adaptability in its doctrine in the form of general principles, based on its perceptions of the United States and other threats to its goals and aspirations. It will seek to avoid types of operations and environments for which U.S. forces are optimized. During the course of conflict, it will make further adaptations, based on experience and opportunity.

When a U.S. force or a U.S.-led coalition first begins to deploy into theater, the OPFOR will seek to disrupt the deployment and thus create opportunity. In such cases, the conventional design the OPFOR used in regionally-focused operations may still provide the framework for military operations against an advanced extraregional force. The OPFOR will not shy away from the use of military means against such an opponent, so long as the risk is commensurate with potential gains. As a U.S. or coalition force builds up power in the region, the OPFOR must rely on adaptive applications of its basic design in order to mitigate its disadvantages and exploit its advantages compared to this new opponent.

In general, the contemporary OPFOR will be less predictable than OPFORs in the past. It will be difficult to template as it adapts and attempts to create opportunity. Its patterns of operation will change as it achieves success or experiences failure. OPFOR doctrine might not change, but its way of operating will.

### INITIATIVE

Like U.S. Army doctrine, OPFOR doctrine must allow sufficient freedom for bold, creative initiative in any situation. OPFOR doctrine is descriptive, but not prescriptive; authoritative, but not authoritarian; definitive, but not dogmatic. The OPFOR that U.S. units encounter in various training venues will not apply this doctrine blindly or unthinkingly, but will use its experience and assessments to interpolate from this baseline in light of specific situations. Thus, U.S. units can no longer say that the OPFOR has to do certain things and cannot do anything that is not expressly prescribed in established OPFOR doctrine. Doctrine guides OPFOR actions in support of the State's objectives; OPFOR leaders apply it with judgment and initiative.

## KEEPING THE COE AND THE OPFOR CONTEMPORARY

The COE is extremely fluid, with rapidly changing regional and global relationships. New actors—both nations and non-state actors—are constantly appearing and disappearing from the scene. The OPFOR doctrinal framework and strategy provided in this manual should meet most of the U.S. Army's training needs for the foreseeable future. During the period covered by the COE, almost anyone who fights the United States would probably have to do it within the general framework described here. As the geopolitical situation, forces, or capabilities change over time, OPFOR doctrine and its applications will evolve along with them, to continue to provide the Army a "contemporary" OPFOR. Thus, the OPFOR will remain capable of presenting a challenge that is appropriate to meet evolving training requirements at any given point in time.

## Chapter 1

# The State

The purpose of this chapter is to introduce the reader to the State. This country is a dominant power in its region of the world. However, it has ambitions of increasing its status in the region and in the global arena. Given the right conditions, the State could challenge the national interests of the United States and its allies. However, the diplomatic-political, informational, economic, and military strengths of the State could be adversely affected by circumstances such as internal strife, economic downturn, or natural disaster. These internal struggles and potential weaknesses within the State could thwart its expansionary goals or even lead to its becoming a failing or failed state in which the government ceases to meet all the needs of the people and at least parts of the country become virtually ungovernable.

Thus, the State is a country in transition. The outcome of that transition is uncertain, and there are a number of possible outcomes that could lead to a number of different scenarios for crisis and conflict. If the State succeeds in its grand ambitions, it could become a major center of global power—either joining the United States and other core states or becoming a competitor. If it maintains the status quo or expands its influence in its region, it could still challenge U.S. interests in that region. If it were to become a failed state, it could present an environment in which the United States and/or other extraregional powers might need to intervene to restore order and deal with humanitarian issues.

This chapter first describes the general characteristics of the State, then the perceived threats to the State's security and national interests. The State regards the United States (or any other wealthy extraregional power) as a potential threat to its regional and global aspirations. Therefore, this chapter also outlines the State's perceptions of the United States, including the State's views of U.S. strengths and vulnerabilities.

## CHARACTERISTICS OF THE STATE

1-1. The State is a nation with a significant geographical area, a growing population, and an emerging economy. Within its region, it is a pivotal nation whose fate or future could affect both regional and international stability. It is not a nation that seeks or maintains close relations with the United States. In fact, the State views the United States as an economic competitor within the region and therefore continually seeks to undermine U.S. relations with other regional nations and members of the international community. The State enjoys great influence within its region, primarily due to its significant military capability and its growing economic base. It is a very aggressive nation and continually analyzes regional and global settings for opportunities that would support its economic expansion and growing diplomatic-political influence. Its Armed Forces are continually modernizing and receive as much as 20 percent of its gross national product (GNP). The State recognizes the dynamic nature of the world environment with its continually changing coalitions, alliances, and partnerships. New actors—both national and transnational—are constantly appearing and disappearing from the scene. The State believes this fluid environment contains many more opportunities than the former Cold-War static international climate afforded. Its long-term goal is to change its position within the global community to one of leadership with a status equal to that of the United States.

## NATURE OF THE GOVERNMENT

1-2. The State is a very powerful, nondemocratic nation that seeks hegemony within its region and the ability to influence the strategic environment in a way that creates conditions favorable to its objectives and economic well being. It is ruled by an elite group of former military officers and wealthy families from a common tribal and cultural background. The President of the State, the nation's senior military leaders, the Minister of Defense, and the Chief of Internal Security are from the same family. The Minister of Foreign Affairs, Minister of Public Information, and Minister of Finance and Economic Affairs are all from the same tribe or cultural group as that family. Each of the State's several district governors is from the ruling class elite and has family or tribal ties to the President. Almost all members of the ruling class have been educated abroad and most have advanced degrees. While there is an elected National Assembly, it has no real power and serves mainly to validate presidential decrees. The State has a legal system and laws based on the religious and cultural standards of the ruling elite.

## REGIONAL AND GLOBAL RELATIONSHIPS

1-3. The State has entered into several strategic partnerships with influential nations, transnational corporations, and transnational organizations. These partnerships serve not only as a means to enhance its security, but also to increase its world economic position. Even though the State is not a democratic nation, its emerging regional influence and increasing economic position have caused several developed and developing states to foster close ties with it. Within its region, its neighbors view the State with concern, primarily because of its aggressive nature and growing military capability. Neighboring states are heavily influenced and intimidated by it. There is also concern because of minority enclaves in neighboring states that share the same cultural, religious, and ethnic background as the State's ruling class and the propensity for these enclaves to look to the State for leadership.

## ECONOMICS

1-4. While not among the world's leading nations economically, the State does maintain a favorable trade balance and possesses a strong industrial base that continues to expand. The State participates in the global market and has been developing an information-age technical base to further its ability to interface with the more developed nations and gain benefit from increased economic globalization. It is the leading economic influence in its region.

## DEMOGRAPHICS

1-5. The population of the State comprises an upper class, a growing middle class, and a very large lower class. The upper or ruling class, as well as the growing middle class, possesses the same ethnic, cultural, and religious foundations. However, the lower class consists of several large minority groups of different ethnic, religious, and cultural backgrounds. The lower-class minorities do not mix and maintain a strong dislike for each other. While they also dislike the ruling class, they are generally supportive of the national government. Most members of the upper class possess university degrees, while most members of the lower class receive vocational training at the high school level. Over 80 percent of the population is literate. Focused public education is directed by the Ministry of Education, in coordination with the Ministry of Finance and Economic Affairs that is charged with ensuring the right work force for future national needs. Even though the population is trained and employed, it is still exploited in order to gain overall economic advantage within the world industrial sector. This leads to lower-class tension and some tension in the middle and upper classes.

## INFORMATION

1-6. The State enjoys tight control over broadcast and print media, and Internet access. The Ministry of Public Information is keenly aware of the role media and information play in international events, diplomatic actions, and military operations. The public has a moderate level of access to open information (radio, television, and news services). The State is an industrial-based society, but it is striving to become an information-based society. It recognizes the value of military and civilian applications of information technology and strives to continuously upgrade its capabilities. Typically, personal computer use has been limited to the elites and members of the military. However, the access to information technology is spreading to the middle class and may present a challenge to the State control of information and its control over Internet access.

## PHYSICAL ENVIRONMENT

1-7. The State has a wide range of topographical features ranging from vast, barren plains to rugged mountains and dense forests. It has several areas that contain large urban centers, and in these areas it has a well established and very robust infrastructure. It also has several wilderness areas that have a very scattered population with an immature road and communications network. The State has several large ports, numerous inland waterways, and a very robust rail network supporting its industrial centers. The climate has extremes of both heat and cold, depending upon the geographic area.

## TECHNOLOGY

1-8. The State's first priority for development or acquisition of technology is to support infrastructure, economic development, and information architecture.

The second priority is for dual-use technology that can serve both the civilian and military sectors. The third priority is uniquely military technologies.

1-9. The State is aggressively developing the industrial and technological base necessary to support indigenous technological research, development, and growth. Meanwhile, it is dependent on foreign resources and imports to sustain many critical research efforts. It invites transnational corporations to establish research and manufacturing facilities in the State as a means of building infrastructure. The State and its wealthy families also own businesses located abroad that can use human and natural resources of other countries to produce desired technology or generate revenues with which the State can buy the technology. The State aggressively seeks to acquire high-technology weapon systems and system upgrades. However, it is aware that it cannot match a technologically advanced opponent and is focusing its efforts on more adaptive and creative uses of technology.

## EXTERNAL ORGANIZATIONS

1-10. The State recognizes that the international environment is a dynamic, changing entity including many types of external organizations that may become involved in regional affairs. For example, humanitarian relief organizations come into the region and the State to deal with natural disasters or the results of civil unrest or armed conflict. These nongovernmental organizations (NGOs) and private volunteer organizations (PVOs) each have their own agendas, which may positively or negatively impact the State. The State attempts to keep out of its territory those external organizations whose interests do not agree with its own. Those that are allowed in are heavily infiltrated and controlled by the State. The State constantly seeks ways to exploit such groups for its own benefit.

1-11. With globalization of economies, the State must also take into account transnational corporations conducting business in the State or its region. If the State's actions adversely affect these foreign enterprises, it could invite outside intervention. However, the presence of foreign business interests and assets in the State may also put additional pressure on an extraregional enemy to avoid collateral damage to civilian life and property if it intervenes there.

1-12. Various transnational groups exist within the State and its region. These groups are often based on demographic, economic, or political issues that transcend national boundaries. Some of these groups are overtly or covertly sponsored by the State. Others the State attempts to infiltrate and manipulate to support its own interests or minimize their negative impact on those interests.

## NATIONAL WILL

1-13. The State projects and promotes a sense of strong unified support for its political leadership, government entities, and its military. It expects and demands such support from its military leaders, elected officials, and general population. The State continuously works to solidify national determination and resolve. Overall, the State's populace possesses a strong willingness to accept hardship and prolonged demands in order to support the agenda of the

State. However, there are factions and groups within the State that do not support this unified position and might, at times, challenge the goals and policies of the State.

1-14. Despite internal differences, the State leadership remains confident that the collective will of the State and its people can give it an advantage over a regional or extraregional enemy. When necessary, it believes it can outlast the opponent's will to continue the conflict. Thus, it is willing and prepared to accept long timelines for strategic campaigns and would seek to use this willingness against a potential opponent.

### TIME

1-15. The State's goals of expanding its influence within the region and the global community are long-term goals. The State is aggressive, but patient. It is willing to spend however much time is necessary to achieve its goals and is satisfied as long as it continues to make progress toward them.

1-16. In the State's view, its long-term goals would be only temporarily thwarted by intervention by an outside force in regional affairs. Accordingly, its planning for dealing with such intervention focuses on effects over time. The State believes that patience is its ally and an enemy of the extraregional force. This gives the State an advantage over an enemy with a short timetable for achieving the goals of his intervention.

### MILITARY CAPABILITIES

1-17. Overall, the State's Armed Forces are well-trained and capable of conducting combined arms and joint operations. Limited to regional power-projection capabilities, the Armed Forces have well-defined acquisition processes and logistics methodologies. The State generally organizes its forces into six service components: the Army, Navy, Air Force, Strategic Forces, Special-Purpose Forces (SPF) Command, and Internal Security Forces. The Army is the dominant partner among the services, but relies on the mobilization of reserve and militia forces to conduct sustained operations. These additional forces are not as well-trained and -equipped as the standing Army.

1-18. The State is beginning to recognize and develop capabilities for the military dimensions of space and information warfare (IW). At this point, however, it has only limited indigenous space capabilities and relies on leveraging commercially available capabilities, while investigating technologies for possible future development.

## PERCEIVED THREATS

1-19. The State faces serious challenges from internal friction, its regional neighbors, and possible intervention by an extraregional power. These threats drive the State's force design and investment strategies.

### INTERNAL

1-20. The State does not have a homogeneous society, but has a variety of religious, ethnic, and cultural factions. Even within the dominating upper class, there can be friction between the State government and the wealthy families. The State may have to deal with terrorist activities, from either internal factions, transnational groups, or international terrorist groups. Drug and criminal organizations within the State or extending into the State can be threats or, under certain circumstances, may become allied with the State's ruling elite.

1-21. As a result of the State's developing a strong industrial base, there is a growing economic gap between "haves" and "have nots" in the State population. The "have nots" of the lower class do not have the same ethnic, cultural, and religious background as the ruling elite or the growing middle class. While the middle class generally supports the State government, some of its members advocate a more democratic form of government, not dominated by the elite of the upper class. Members of the lower class also generally support the State government, partly because most of them work in the factories owned by a few wealthy families, who also control the State government.

1-22. However, many in the lower and middle classes have some mistrust of the government because of the ruling class's alliances with other nations and transnational organizations. In particular, they feel betrayed by this outside influence, from which primarily the upper class profits. Without real voting power, the lower and middle classes believe that the primary tools they can use to influence the government are street riots, demonstrations, and work stoppages.

1-23. The more disaffected members of the middle and lower classes have from time to time attempted to launch insurgent movements. However, these antigovernment groups are fragmented due to ethnic, cultural, and religious differences between the middle and lower classes and the mutual dislike among the various large minority groups in the lower class. The most successful insurgent groups receive backing from populations of neighboring countries with similar ethnic, religious, or cultural heritage.

1-24. There is a growing trade in illegal narcotics. While the State government officially condemns the drug trafficking, the wealthy families of the ruling elite actually control and profit from the narcotics business. Members of the lower class (particularly in rural areas lacking factory jobs) rely on the drug organizations for their livelihoods. The drug organizations' operations are not confined to the State, where some such organizations enjoy unofficial government protection. A drug organization can protect its personnel and assets by maintaining its own paramilitary and security forces, outfitted with modern arms purchased with drug money.

## REGIONAL

1-25. The State develops all its instruments of national power with the goal of regional dominance. For example, it designs its military forces and employs an investment strategy in military materiel that ensures its Armed Forces conventional power superiority over any of its regional neighbors—individually, if not collectively.

1-26. Rarely are only two sides involved in conflict. It is far more likely that a coalition of actors with similar interests opposes another coalition or other actors. The regional environment can also include neutral parties and subnational or transnational actors with interests that support or conflict with those of the State.

1-27. The State can deter other regional actors from actions hostile to the State's interests by maintaining a trained and ready military force. When necessary, the State conducts conventional operations that form the core of its ability to dominate regional adversaries or to compel such adversaries to yield to the State's will. Military force, or its threatened use, is a key element of the State's status as a power within the region. However, the State seldom applies military power alone; such power is most effective when applied in combination with informational, economic, and diplomatic-political means.

## EXTRAREGIONAL

1-28. The State sees intervention by an extraregional power (such as the United States) as a threat to its aspirations and influence within the region. It fears that international sanctions may set back its economic growth and regional influence. Therefore, it attempts to keep any regional conflict below the threshold that would invite outside intervention. Nevertheless, an extraregional power may find it necessary to intervene in regional affairs. It may do so unilaterally or, more likely, as part of a coalition under United Nations mandate.

1-29. In order to deal with extraregional intervention, the State will use all means necessary (diplomatic-political, informational, economic, and military). The same investment strategy that gives the State military superiority over its regional neighbors can assist it in deterring an extraregional power from entering the region or dealing with that power if intervention does occur. The State must go to war or continue the war after extraregional intervention with whatever means it has. However, it must use those means in adaptive ways.

1-30. Realizing that its regional aspirations and actions may trigger extraregional intervention, the State may invest in certain high-payoff technological niches that offset the advantages of an extraregional power. Examples could be investment in a more robust air defense capability than would be necessary to deal with air threats from regional adversaries or in GPS jammers that are effective only against a force relying on GPS. However, most investments will address both regional and extraregional threats. The basic force structure of the Armed Forces is the same for either type of threat.

1-31. Once an extraregional force deploys into the region, the State uses an integrated aggregate of all the forces and means that were already available to it for internal and regional threats. For example, Internal Security Forces take firmer control of the population because of the impending attack on the homeland. These security forces may also be called upon to fight enemy forces while simultaneously trying to preserve the regime and the instruments of national power.

## VIEWS OF THE UNITED STATES

1-32. The State constructs its military capabilities in accordance with its assessments of internal, regional, and extraregional threats or opportunities. In evaluating extraregional threats, the State bases its estimates upon its perceptions of the United States and other modern wealthy nations.

### OVERALL ADVANTAGE IN WARFIGHTING CAPABILITY

1-33. The State generally sees the United States as the sole superpower, with an overall advantage in technology and warfighting capability. Today's perceptions of combat operations have been filtered through the lens of the Gulf War, Bosnia, and Kosovo. These perceptions include U.S. forces conducting deliberate actions at a tempo decided by them. Such operations are characterized by the application of technology and advanced systems that leave opponents virtually helpless to respond or retaliate by traditional means. Despite these strengths, however, the State sees some weaknesses that adversaries of the United States may be able to exploit.

### VULNERABILITY OF COALITIONS

1-34. In extraregional intervention, the United States usually acts within the confines of a political and/or military coalition. This is particularly true in today's world of economic interdependence and political interaction that require a nation's military forces to coordinate and work with allies, NGOs, and other governmental agencies and services during the conduct of operations. Establishing and maintaining command and control ($C^2$) of all the players may prove difficult. Compared to a long-term alliance, a coalition formed for a specific purpose is likely to have problems with interoperability, language, and lack of a common operational framework. There may also be mistrust and problems in sharing classified information.

1-35. A coalition is normally only as strong as its weakest member. The State understands the weakness of coalitions and alliances and will seek to force its regional neighbors or the United States to create alliances with nations who are more sympathetic to its own cause. It will try to create or highlight differences among coalition members and use this to cause a split or to hamper coalition objectives.

### UNWILLINGNESS TO ACCEPT HEAVY LOSSES

1-36. The United States is unwilling to accept heavy losses and is risk-averse. The American people do not like to see their soldiers getting killed in other countries, especially if they do not think it is vital to U.S. interests. During the war in Vietnam, television brought the war home, into American living rooms, and the U.S. public did not like what it saw. More recently, in Somalia, U.S. opponents dragged the body a downed U.S. helicopter pilot through the streets and were thus able to destroy U.S. public support for a continued military presence there. The State will try to inflict highly visible and embarrassing losses on U.S. forces to weaken U.S. domestic resolve and political will to continue a conflict. National will is universally perceived as a strategic center of gravity for the United States.

1-37. Technologically advanced societies count on quick, easy victory. They are generally hard-pressed to conduct a protracted, high-casualty war of attrition. Thus, the U.S. Armed Forces seek a quick war of annihilation with an overmatched foe. U.S. public perception of current combat operations is based on a premise of low casualties, a secure homeland, precision attacks, and relatively short duration of conflict.

## SENSITIVITY TO PUBLIC OPINION AND LACK OF COMMITMENT

1-38. The U.S. political leadership is very sensitive to the opinion of its own people and to world opinion. Since the Vietnam War, other nations have perceived that the United States lacks commitment over time. Public opinion can prevent or delay U.S. involvement in a crisis, cause loss of effectiveness during involvement, or cause U.S. withdrawal from previous commitments. This perception impacts U.S. relations with its allies, and the State will try to use it against the United States.

## PREFERENCE FOR STANDOFF COMBAT

1-39. The United States avoids close combat and relies on air campaigns and standoff technology with long-range, precision weapons. It demonstrated this tendency in Desert Storm, Bosnia, and Kosovo. Often, U.S. ground forces are used to support their air forces by guarding their airfields and holding an opponent at bay while the U.S. Air Force attacks targets. The United States prefers to use prolonged air campaigns, missile attacks, and information attacks before committing ground forces to ensure a quick victory.

## LACK OF OPTIMIZATION FOR CLOSE, DISMOUNTED COMBAT

1-40. U.S. ground forces prefer to fight the long-range battle and are reluctant to engage in close combat. The U.S. Army lacks sufficient dismounted infantry forces for combat in complex terrain (cities, forest, mountains, jungle, or swamp), where close combat is likely to predominate. Combat in such terrain swings the advantage from the side with the more advanced technology to the side that has the most trained infantry and is willing to fight a protracted war. The best places to counter U. S. forces are in cities and other complex terrain, as well as in other locations where the U.S. must conduct close combat operations.

1-41. U.S. mechanized forces are reluctant to fight separated from their vehicles. Personnel carriers are left in a "follow and support" mode and are seldom consolidated to form a mobile, armored reserve. This reluctance keeps U.S. mechanized forces rather "road bound."

## DEPENDENCE ON HIGH TECHNOLOGY

1-42. U.S. military operations depend on high-technology equipment working flawlessly. However, it is possible for any technology to be defeated by matching or lesser technology. For every system there is a counter—often a simpler and less expensive counter. High-technology equipment cannot perform optimally over time without extensive down time for maintenance. A system's technological advantage drops off dramatically with the onset of combat. Over time, older, simpler, and more rugged systems often outperform complicated, high-technology systems that are less field-worthy and require extensive maintenance.

1-43. Key components of U.S. military thinking include precision strike missions, automated $C^2$, information operations, and electronic strike missions. The Gulf War provided indications as to how high-technology permeated U.S. mission planning and execution. Subsequent conflicts have highlighted ways that State forces can adapt to offset U.S. technological advantages.

## DEPENDENCE ON INFORMATION DOMINANCE

1-44. The United States is trying to build forces that use information dominance in order to execute precision strikes and maneuver. In particular, U.S. forces are coming to rely on advanced $C^2$; computers; and reconnaissance, intelligence, surveillance, and target acquisition (RISTA) technology interfaced with higher headquarters and outside agencies. This technology and the necessary communications and data links are critical to maintaining enhanced situational awareness. However, forces relying on such capabilities can suffer from information denial or information overload.

1-45. The very systems and links upon which U.S. forces rely are also high-payoff targets for computer warfare, information attack (IA), or physical destruction. Denial of these resources at critical times can deny U.S. forces complete situational awareness. The State can deny U.S. forces situational awareness with little investment in terms of time, assets, or infrastructure by using common jamming systems and other off-the-shelf technologies.

1-46. U.S. units are very susceptible to media influence and information manipulation due to their robust array of sophisticated RISTA systems and widespread access to information. Large numbers of U.S. sensors can overwhelm their own units' ability to receive, process, and analyze raw intelligence data and to provide timely and accurate intelligence analysis. Thus, it is not necessary to destroy or disrupt the U.S. collection capability. The State can also overload U.S. analysis capability by saturating U.S. sensors with real and/or deceptive data.

1-47. Another perception is that U.S. ground forces do poorly in counterreconnaissance planning and execution. The State typically places high priority on the identification, location, destruction, disruption, or deception of U.S. reconnaissance elements. It can use a high volume of counterreconnaissance, sentries, and observers, as well as the indigenous population, to report on the presence and activities of U.S. elements. It can also use urban areas and other complex terrain to limit the effectiveness of sophisticated RISTA assets and deny the U.S. an advantage in situational awareness.

## PREDICTABLE OPERATIONS

1-48. The U.S. military conducts operations that are rather predictable and templatable. This can allow the State to avoid the intended effects of U.S. operations and to maximize the effectiveness of its own forces and weapons against U.S. forces. U.S. planners often apply templates from one region or area to another without considering the differences in terrain, economic development, or social culture. To a certain extent, U.S. reliance on technology contributes to predictability of U.S. operations.

1-49. U.S. soldiers and noncommissioned officers (NCOs) are well trained and show a great deal of initiative, aggressiveness, and self-reliance. U.S. officers, however, may be less aggressive and are tied to a chain of command that often takes over the fight from its company- and field-grade officers. U.S. field-grade officers have difficulty coordinating their diverse combat elements in order to apply maximum combat power in a timely manner.

## LACK OF CULTURAL AWARENESS

1-50. The U.S. armed forces generally are culturally unaware of many of the regions of the world in which they may be committed and fail to understand fully the issues and peoples that confront them. U.S. planners do not give adequate consideration to cultural differences in various operational environments.

## VULNERABILITY OF FORCE PROJECTION

1-51. U.S. ground forces require a long time to deploy into a theater and develop it before they can conduct effective combat. The United States has failed to develop sufficient transport for force projection of a potent combat force. Therefore, the U.S. Army initial-deployment forces are lightly armed and equipped with limited firepower, limited logistics support, and limited communications. They can land only at ports, harbors, and airfields. U.S. forces are not designed for forced entry, but do have some forced-entry capability. However, they have limited staying power.

## DEPENDENCE ON ROBUST LOGISTICS

1-52. U.S. forces are dependent on an extraordinarily complex and comprehensive logistics system. A large percentage of U.S. forces is tied up in logistics, since the U.S. military personnel require far more supplies and creature comforts than other armies do. The U.S. forces usually prefer to deploy into prepared theaters backed by forward-deployed forces and logistics bases. The operational environment may not always permit such preparation or possess local infrastructure that can support U.S. operations. The United States is trying to address this problem with the use of high technology, just-in-time logistics, and contractor support, which lead to vulnerabilities of their own.

1-53. U.S. forces may have insufficient organic logistics force structure to adequately sustain resource-intensive operations in urban areas and/or other complex terrain. Thus, it may be easy for the State to destroy or disrupt U.S. sustainment operations. The State's bypassed forces can isolate U.S. units from their combat service support. The State can increase the strain on U.S. logistics by temporarily closing air or seaports. It can orchestrate displaced civilians or instigate riots and demonstrations to block or congest U.S. resupply routes. If the State can intimidate local nationals or turn them against the U.S. force, loss of host nation support could place a greater strain on the already limited U.S. sustainment assets.

## RELIANCE ON CONTRACTOR SUPPORT

1-54. Once in a theater, the U.S. forces are dependent on a large number of civilian contractors, both U.S. and local, to maintain their equipment and perform a number of essential tasks. Such contractors are vulnerable targets for hostile forces. U.S. forces may have to provide force protection for contractor personnel, thus decreasing combat power. Threat conditions may increase to a level where contractors can no longer be used, placing greater strain on limited U.S. organic sustainment assets. Contractors and the host nation network through which they must work increase the vulnerability for information exploitation and sabotage.

## DOWNSIZING AFTER CONFLICT

1-55. Historically, the United States has often downsized its military forces after conflicts (WWI, WWII, Vietnam, and the Cold War) to the extent that remaining forces were ill-prepared and -equipped to conduct their missions. The United States usually loses its first ground battle following severe cuts. Although its resources are stretched, the United States still pursues a policy of increasing involvement worldwide. Therefore, it may not be able to bring sufficient forces to bear in a timely manner to exploit its technological overmatch against the State in the State's region of the world.

# Chapter 2

# Structure of the State

The State's ruling elite understands that the power the State derives from maximizing its strengths in order to compensate for its weaknesses. The State judiciously uses its elements of power to influence the behavior of other states and to exercise control over them. The State is the dominant actor in the region, and it intends to remain so and continue to expand its influence over neighboring states. In order to assist it in achieving its expansionary goal, it has established a solid infrastructure that allows it to develop and implement a policy designed to maximize the benefits it may gain from its tangible and intangible resources. This chapter introduces the reader to the elements of power and explains how the State can and will use these elements to its advantage. Then, this chapter outlines the State's organizational structure which, combined with the State's use of the elements of power, allows it to be the most powerful nation in the region.

## ELEMENTS OF NATIONAL POWER

2-1. The State defines *national power* as one nation's capacity to influence the behavior of other nations or non-state actors or exercise control over them. Power may be real or perceived, tangible or intangible. The distinction between perceived and real power is the art of statecraft. When states perceive that a certain state is powerful, that perception becomes reality. The "powerful" state exerts control and influence over the states that hold the perception. The elements of national power are divided into two interrelated categories: sources and instruments of power.

### SOURCES OF NATIONAL POWER

2-2. Many of the characteristics of the State described in Chapter 1 are sources of power. The State strives to transform its tangible and intangible resources into useable instruments that can increase its status in the international community. Sources of power include—

- Geography.
- Population.
- Economy.
- National will.
- National direction.

These sources of power are not directly employable, but they provide conditions for generating and employing the State's instruments of power.

### Geography

2-3. Geography sets the conditions and provides the resources for the generation of power. The State uses its geography and physical infrastructure to its advantage. Its natural ports and inland waterways, coupled with its extensive rail network, allow it to import those goods and services that it does not produce in its industrial centers but are necessary for the conduct and sustainment of wars. The State can nationalize all public and private transportation networks in order to exert the maximum amount of control over them.

### Population

2-4. As the State grows in population, it may have a manpower advantage over a less populous neighboring country. Its people are industrious and creative and generally support the State government. Support of the populace enables the State to exert its power, influence, and control within its own boundaries and within its region. The State believes that collective will of the people can give it an advantage over a regional or extraregional opponent.

### Economy

2-5. A nation's economic level directly affects that nation's ability to operate competitively in an international environment. The State takes advantage of its strong economic base to develop a favorable balance of trade. Its strong, expanding industrial base allows it to conduct sustained military operations. By participating in the global market and fostering transnational ties, the State has developed favorable relationships with critical trading partners who can help support it during peacetime and during times of war.

### National Will

2-6. National will is an intangible but very important source of national power. As a nondemocratic nation, the State demands hegemony within its borders. An elite group of former military officers and wealthy families who share the same tribal or cultural background control the State's government. The State manipulates its people's ideas and values, promoting those that agree with the government's and suppressing those that do not. It demands total support from its populace and will not tolerate antigovernment sentiment. The populace is forced to support State policies.

### National Direction

2-7. The State's government and military provide unity of effort and develop a clear vision for the State's future policy goals, commitments, and programs. The degree of centralized government control, coupled with the small size of the ruling elite, allows the State to unite all organizations and bureaucracies to support its common aims. In times of war or crisis, the State can task its Internal Security Forces to monitor potentially troublesome groups or factions and to rally the political support of the general population.

## INSTRUMENTS OF NATIONAL POWER

2-8. Instruments of national power represent tangible resources that can be manipulated, altered, and balanced. The State recognizes the following instruments of national power, which are interrelated:

- Diplomatic-political.
- Informational.
- Economic.
- Military.

These instruments of power are complementary, and the State employs them in varying combinations as components of its overall national security strategy.

2-9. The State clearly understands the importance of meeting the objectives of its national security strategy. It intends to achieve these objectives through the integration and use of diplomatic-political, informational, economic, and military instruments of power. A particular instrument of power may be more dominant or visible given the situation; however, the State will use all instruments to support its strategy and goals and achieve victory.

2-10. The State believes that a purely military strategy for a nation is no longer possible. A clear-cut line of demarcation between military, economic, and political matters does not exist. The informational element cuts across the other three. Power is a combination of many elements.

### Diplomatic-Political

2-11. The State has a diplomatic corps that promotes its goals and protects its national interests within the international environment. These diplomatic means support its national security strategy regionally and internationally. The State utilizes all possible international or diplomatic channels to implement its foreign policy or to support any conflicts. When possible, it establishes alliances and coalitions to assist it in meeting its goals. It maximizes its use of negotiations, recognition of new governments, treaties, and alliances.

2-12. While diplomatic power applies to external relations, the political aspect is internal to the State. The State tasks all internal government agencies to support the national security strategy. In times of war or crisis, it uses these agencies to convince the population to follow its policies.

### Informational

2-13. The State communicates and controls information in order to inform its own population and foreign nations about its policies and actions and to create a favorable response. By implementing a well-organized internal media campaign and information warfare (IW) effort, the State can control dissemination of all information within its borders. It can control and manipulate the content and flow of international information as well as domestic information. Since the control of all information is critical to the State, all State agencies and departments follow strict guidelines to ensure the total control and appropriate dissemination of information.

and military relations; and attempts to foster support and cooperation from other nations. In particular, the Ministry of Foreign Affairs cultivates useful relationships with other nations in the region. It also strives for recognition of the State as a participant in global affairs. (See Figure 2-2).

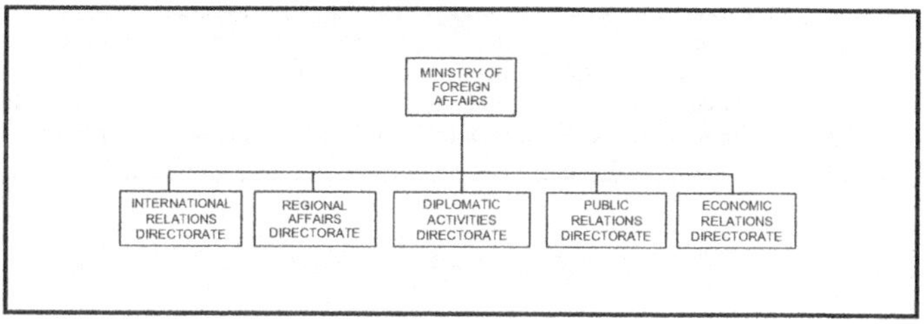


**Figure 2-2. Ministry of Foreign Affairs**

## INTERNATIONAL RELATIONS DIRECTORATE

2-23. The International Relations Directorate is responsible for fostering positive relations with regional and extraregional countries. It establishes cultural and educational exchanges with these countries in order to promote these positive relationships. Exchanges may include but are not limited to sporting events, educational aid to international students, traveling lecture series, and performing arts shows. The International Relations Directorate is not responsible for formulating diplomatic policy.

## REGIONAL AFFAIRS DIRECTORATE

2-24. The Regional Affairs Directorate represents the State at regional conferences, meetings, and political events. It attempts to foster support for the State and cooperation from other regional actors.

## DIPLOMATIC ACTIVITIES DIRECTORATE

2-25. The Diplomatic Activities Directorate is responsible for promoting diplomatic relations within the region as well as external to the region. The directorate establishes diplomatic ties, negotiates treaties, and attempts to extend the political influence of the State.

## PUBLIC RELATIONS DIRECTORATE

2-26. The Public Relations Directorate conducts information campaigns (called public relations campaigns) within the region and external to the region. It produces public relations films and articles and coordinates events that extol the virtues of the State.

## ECOMONIC RELATIONS DIRECTORATE

2-27. The Economic Relations Directorate promotes the economic policy of the State. Constantly looking for ways to improve the economic base of the State, this directorate attempts to expand State markets and find procurement sources for those materials the State cannot or does not produce. The directorate's goal is to ensure that the State remains the most powerful in the region economically.

## MINISTRY OF PUBLIC INFORMATION

2-28. The Ministry of Public Information is responsible for the control and dissemination of all political, diplomatic, economic, and military information to the State's populace, other regional actors, and the world. Information is rigidly controlled to support the State. The ministry is also a key player in the development and execution of all strategic IW campaigns. (See Figure 2-3.)

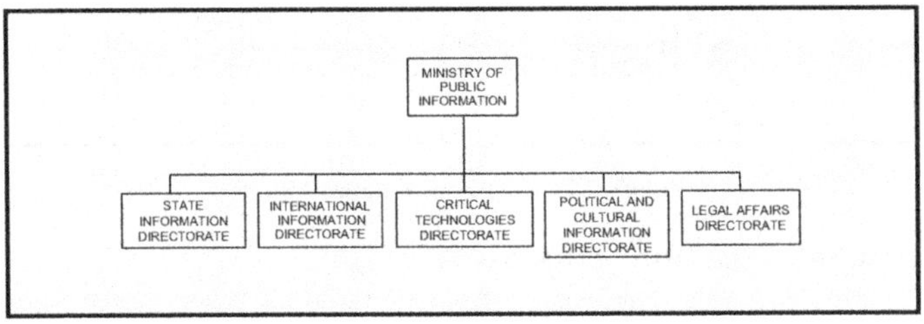


**Figure 2-3. Ministry of the Public Information**

### STATE INFORMATION DIRECTORATE

2-29. The State Information Directorate is responsible for the control and dissemination of information within the State. This directorate coordinates with the SID to encourage public support for the State's policies and programs and to counter hostile attempts to distort and frustrate those efforts. It monitors and controls all internal media sources and reviews and censors all information prior to authorizing its public release. With the assistance of the State's internal security forces, it tracks down sources of dissident or underground newspapers and, upon finding them, halts publication. When possible, it tries to focus known internal subversive elements in a direction that the regime can control.

### INTERNATIONAL INFORMATION DIRECTORATE

2-30. The International Information Directorate maintains tight control of information released to the international world. This directorate attempts to manipulate the international media and the consumers of this information to ensure that they will actively support the policies and goals of the State. Thus, it plays a major role in perception management activities to influence international political and public opinion regarding the State and its actions.

### CRITICAL TECHNOLOGIES DIRECTORATE

2-31. The mission of the Critical Technologies Directorate is twofold. First, it filters and protects the amount of information it releases to the State public and to the world on the technological capabilities of the State. Second, it consistently researches ways to acquire new technologies to improve the State's technological base.

### POLITICAL AND CULTURAL INFORMATION DIRECTORATE

2-32. The Political and Cultural Information Directorate plays a primary role in producing government-approved information on political and cultural matters and disseminating it to the all the State's residents. This directorate coordinates with the Ministry of Education to disseminate such information through the public schools. Other recipients of this information campaign include the youths who are required to join the State Youth Corps, the workers who work within the State's industrial and agricultural base, and all other residents including those who serve at the highest military and governmental levels.

### LEGAL AFFAIRS DIRECTORATE

2-33. The Legal Affairs Directorate provides legal advice and assistance to the State. In wartime, this directorate works in close coordination with the Legal Affairs Department of the Logistics Directorate in the Ministry of Defense.

## MINISTRY OF FINANCE AND ECONOMIC AFFAIRS

2-34. The Ministry of Finance and Economic Affairs formulates, implements, and promotes the economic policy objectives of the State. It represents the State in economic negotiations with foreign governments and international organizations, explains its economic position and views, endeavors to promote its economic relations, and attempts to foster economic support and cooperation from other nations. (See Figure 2-4.)

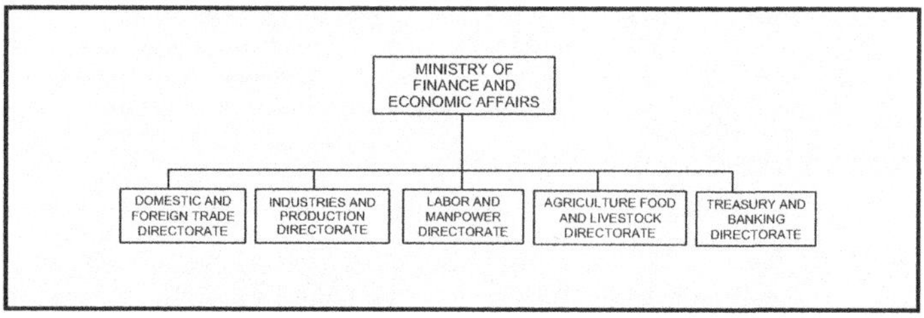


**Figure 2-4. Ministry of Finance and Economic Affairs**

### DOMESTIC AND FOREIGN TRADE DIRECTORATE

2-35. The Domestic and Foreign Trade Directorate continually seeks ways to improve the State's balance of trade. It watches domestic production and consumption figures to determine where surpluses exist that may be exported in order to increase the State's economic revenue. It exercises complete control over domestic production, and it directs what will be exported and to what country. The directorate sends representatives to regional and extraregional countries to seek foreign imports and to arrange for foreign exports. All transactions are based on what will benefit the State. This directorate manages an active program for protection of sensitive economic information related to production, imports, and exports. However, it also executes a carefully orchestrated economic information campaign for effective dissemination of appropriate information on the economic situation and economic policy, both within the State and externally.

## INDUSTRIES AND PRODUCTION DIRECTORATE

2-36. The Industries and Production Directorate controls all aspects of industrial production within the State. It determines what will be produced and by whom. It establishes strict quotas for the factories. Should factories fail to meet these quotas, the directorate determines who is responsible for the failure, and that person is severely punished. The directorate establishes extensive outyear plans that will produce an industrial environment that continues to support the State as it expands its economic status within the global community.

## LABOR AND MANPOWER DIRCTORATE

2-37. The Labor and Manpower Directorate controls all aspects of the State's work force. The directorate determines where labor is needed and assigns personnel to work in those areas. During their intermediate school years, all students are given State-developed aptitude tests. The State then assigns these students to specialized secondary schools that will train them to perform in whatever career for which the State deems they are suited.

## AGRICULTURE AND LIVESTOCK DIRECTORATE

2-38. The Agriculture and Livestock Directorate directs all aspects of agriculture and livestock production. It assigns production quotas to individual farmers. It also runs State-owned collective farms to which it also assigns mandatory quotas for production. The directorate involves itself in technological research in order to find ways to improve production.

## TREASURY AND BANKING DIRECTORATE

2-39. The Treasury and Banking Directorate controls all monetary assets and fiscal matters within the State. Fiscal matters include but are not limited to: the cost of living index, the inflation index, the monetary exchange rate, the balance of payments, and interest rates. The directorate conducts regular reviews of State fiscal policies to ensure that the State receives maximum advantage from its policies.

# MINISTRY OF THE INTERIOR

2-40. The State, like most nondemocratic nations, maintains large internal security forces to deal with various internal threats to the regime. In peacetime, the Chief of Internal Security heads the forces within the Ministry of the Interior that fall under the general label of "internal security forces." (See Figure 2-5.) These forces are responsible for internal security and all related functions. Members of these forces are selected from segments of the population most loyal to the State government. (See FM 7-100.3 for more detail on the various types of internal security forces.)

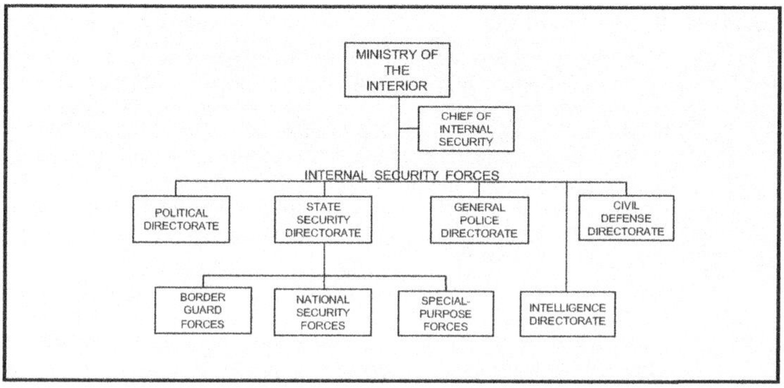


**Figure 2-5. Ministry of the Interior**

2-41. Internal threats do not exist only in peacetime, but also continue and often intensify during war. For instance, a regional enemy sharing a common ethnic, religious, or cultural heritage with segments of the State's population may incite or support anti-State activities by those groups. During wartime, therefore, some or all of the internal security forces from the Ministry of the Interior become subordinate to the Supreme High Command (SHC). At that time, the formal name "Internal Security Forces" applies to all forces resubordinated from the Ministry of the Interior to the SHC, and the General Staff controls and supervises their activities. In war, the Internal Security Forces pick up additional missions that support the State in dealing with external elements.

2-42. The primary imperative for the State is preservation of the regime and all four instruments of national power. The State will use all means available to—

- Protect political centers and the political leadership.
- Protect and control information.
- Protect key economic centers.
- Protect military forces.

2-43. Internal security forces aggressively suppress or crush organized groups of dissidents, using force when necessary, regardless of whether the dissident actions are violent or peaceful. The government may even use Armed Forces against such groups. Government-controlled media either do not report the incidents or manage public perception to put the blame on the antigovernment group.

## POLITICAL DIRECTORATE

2-44. The Political Directorate monitors the political activities of the State's population. It infiltrates possible subversive groups and tries to undermine groups' cohesiveness or redirect their efforts in directions the regime can control.

## STATE SECURITY DIRECTORATE

2-45. The State Security Directorate is responsible for preventing antigovernment activities, investigating these activities, and prosecuting the perpetrators. During times of crisis and wartime, the directorate is also responsible for finding and neutralizing dissidents and spies. Elements of the directorate deploy throughout the State. Many of these elements are paramilitary units equipped for combat. They include Border Guard Forces, National Security Forces, and Special-Purpose Forces. Together with the regular Armed Forces, these paramilitary forces help maintain the State's control over its population in peace and war.

2-46. The Border Guard Forces patrol the State's borders, both land and water. They maintain security against unauthorized crossings into or out of the State. They are charged with detecting, identifying, and intercepting illegal infiltrations and with arresting anyone attempting to exit the country unlawfully.

2-47. The National Security Forces are organized along military lines and equipped with light weapons and sometimes heavy weapons and armored vehicles. These units focus on subversive elements and crimes against the government. However, the State will also use these forces against regional or extraregional enemies that invade the State. In turn, it can use its Armed Forces against internal adversaries, when necessary.

2-48. Within the State Security Directorate, the Ministry of the Interior has its own Special-Purpose Forces (SPF). These commando-type forces typically operate in small SPF teams and are the most highly-trained and best-equipped of the internal security forces. They are multifaceted, but are primarily used for VIP security, hostage rescue, counterdrug, counterinsurgency, and counterterrorist activities. They also have the capability to conduct direct action attacks, such as sabotage in the enemy's rear area. They can infiltrate and disrupt enemy organizations (political, social, religious, and military). They can also engage in intimidation, extortion, atrocities, kidnapping, and assassination.

## GENERAL POLICE DIRECTORATE

2-49. The General Police Directorate has responsibility for national, district, and local police, fire protection, and the penal system. Its forces include the paramilitary National Police and special police units that are equipped for combat. The National Police focus almost entirely on maintaining internal security; they are charged with protecting government facilities and with suppressing dissidents. Special police units assume responsibility for crushing organized groups of dissidents and for domestic counterintelligence. Because some special police units are equipped with heavy weapons and armored vehicles, they do provide combat potential to conduct defensive operations if required.

### CIVIL DEFENSE DIRECTORATE

2-50. The Civil Defense Directorate comprises a variety of paramilitary and nonmilitary units. These units have the collective mission of protecting the population and economic centers against the effects of all types of natural disaster and warfare, as directed by the SID. In peacetime, normal missions include emergency engineering, rescue, and similar disaster relief functions. During wartime, civil defense units focus on identifying and repairing battle-damaged facilities and structures including roads, bridges, airfields, and depots. They sometimes protect important military, political, or economic centers against sabotage by internal or external threats and could also perform other support-zone security missions.

### INTELLIGENCE DIRECTORATE

2-51. The Intelligence Directorate is responsible for identifying and neutralizing subversive elements, as well as unsanctioned drug and criminal organizations. It investigates and monitors subversive groups and infiltrates their ranks. It routinely eavesdrops on communications of such groups and their known leaders. It employs an extensive human intelligence (HUMINT) network. It also monitors and collects information on foreign organizations operating within the State. These include not only foreign spies, but also foreign-based corporations doing business in the State, as well as nongovernmental and private volunteer organizations that government agents may infiltrate. The State must examine such external organizations to determine how to deal with negative influences or how it can manipulate these organizations to support its own national security objectives.

2-52. In wartime, the assets of the Intelligence Directorate support the Internal Security Forces and the overall national security strategy. The Internal Security Forces continue to operate an extensive HUMINT network within the State as the cornerstone of its reconnaissance, intelligence, surveillance, and target acquisition (RISTA) capability. During regional conflict, the State expands the HUMINT network into enemy territory. This HUMINT network will also help provide the State a RISTA capability that may offset the technological advantages of an extraregional force.

## MINISTRY OF DEFENSE

2-53. The NCA exercises control over the makeup and actions of the Armed Forces through the Ministry of Defense (MOD) and the General Staff. The MOD has six directorates. (See Figure 2-6.) It is responsible for the day-to-day administration of the Armed Forces and for the readiness and overall development of the six service components of the Armed Forces. However, the General Staff has direct control over the six services. The General Staff also oversees three functional directorates. In wartime, the MOD and General Staff together form the Supreme High Command (SHC). (See Figure 2-7.)

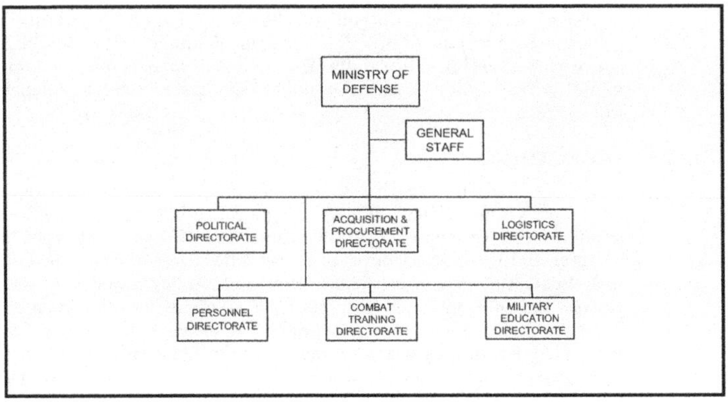


**Figure 2-6. Ministry of Defense**

## POLITICAL DIRECTORATE

2-54. The Political Directorate maintains political control over the Armed Forces and ensures that they adhere to State government policies and directives. It directs the political education, indoctrination, and political activities of all Armed Forces personnel. It maintains records on possible subversive members of the Armed Forces and monitors them to ensure that they remain loyal to the State and do not initiate any actions contrary to State interests.

## ACQUISITION AND PROCUREMENT DIRECTORATE

2-55. The Acquisition and Procurement Directorate coordinates the procurement of all ammunition, weapons, and military equipment from domestic and foreign sources. It ensures that the Armed Forces have sufficient quantities of the types of equipment necessary to achieve State goals. See Chapter 7 for more detail on the functions of this directorate.

## LOGISTICS DIRECTORATE

2-56. The Logistics Directorate is responsible for the overall logistics support of the Armed Forces, movement of troops and supplies, security of logistics installations, maintenance of lines of communication (LOCs), and general defense of the support zone. See Chapter 7 for more detail on the functions of this directorate.

## PERSONNEL DIRECTORATE

2-57. The Personnel Directorate formulates and implements national-level military personnel policies, allocates personnel among the service components, and establishes basic personnel administration procedures. See Chapter 7 for more detail on the functions of this directorate.

## COMBAT TRAINING DIRECTORATE

2-58. The Combat Training Directorate plans, coordinates and supervises combat training in all service components. It evaluates the state of preparedness and combat efficiency of the Armed Forces and makes recommendations to the State for improvement. See Chapter 7 for more detail on the functions of this directorate.

## MILITARY EDUCATION DIRECTORATE

2-59. The Military Education Directorate supervises and coordinates the overall military school system. It provides guidance for curriculum development and ensures that the school system teaches relevant and pertinent information. Direct control of service schools and their curricula rests with the service components. See Chapter 7 for more detail on the functions of this directorate.

# SUPREME HIGH COMMAND

2-60. While the Armed Forces are not a unified defense force, they do have clear lines of command and control ($C^2$). The State's NCA exercises $C^2$ via the SHC, which includes the MOD and a General Staff drawn from all the service components. (See Figure 2-7.) In peacetime, the MOD and General Staff operate closely but separately. The MOD is responsible for policy, acquisitions, and financing the Armed Forces. The General Staff promulgates policy and supervises the service components; its functional directorates are responsible for key aspects of defense planning. During wartime, the MOD and General Staff merge to form the SHC, which functions as a unified headquarters.

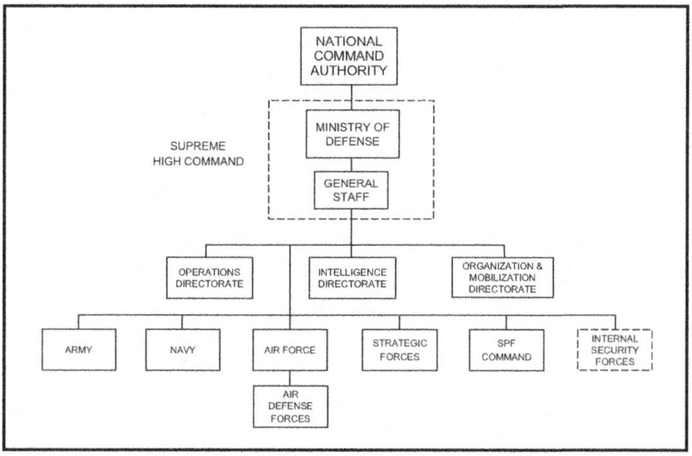


**Figure 2-7. Supreme High Command**

## FUNCTIONAL DIRECTORATES

2-61. Under the General Staff are three functional directorates for operations, intelligence, and organization and mobilization. These directorates handle functions that transcend service boundaries.

### Operations Directorate

2-62. The Operations Directorate is actually focused on *developing plans* for the General Staff or SHC, rather than supervising execution of national-level operations. In fact, the Operations Directorate is the strategic planning organization for the Armed Forces. During peacetime, it is responsible for the development, staffing, promulgation, and continuing review of Armed Forces mobilization and strategic campaign plans. It also reviews mobilization plans of service components and important government ministries for the General Staff and the NCA. The Operations Directorate plays the lead role in developing the national exercise program and provides support to General Staff inspectors assigned to assess service performance in national exercises.

2-63. During combat operations, the Operations Directorate is responsible, with the Intelligence Directorate, for maintaining a continuous estimate of the situation for the SHC, and it contributes to the formal national estimate process. The Operations Directorate also modifies or develops military strategic campaign plans based on guidance from the SHC. The directorate is used to generate options and focuses on the long-, mid-, or short-term objectives of the State. Consequently, the Operations Directorate assigns liaison officers to important government ministries.

### Intelligence Directorate

2-64. The Intelligence Directorate fields national-level reconnaissance assets and supports national intelligence requirements. At this level, limited investments in space-based technology and larger investments in buying products from other nations produce the greatest overall payoff. The State is able to use commercial enterprises of other countries to obtain space-based imagery intelligence and to support command and control.

2-65. The Armed Forces place a premium on HUMINT sources. They operate HUMINT agents throughout the region and have some global capability against those nations the State perceives as the most likely potential enemies. The SPF Command, with both a regional and global capability, also supports national intelligence requirements.

2-66. The State can procure signals intelligence (communications and electronic intelligence) means on the international market. Accordingly, the Armed Forces' capability is mixed and probably difficult to maintain, but not inconsequential. Although the State has targeted regional opponents, it does have some capability against the bandwidths in which it expects likely extraregional opponents to operate. The Army, Navy, and Air Force components have tactical assets appropriate to their respective ground, sea, and air venues.

2-67. The Intelligence Directorate is responsible for acquisition of unmanned aerial vehicles (UAVs, including drones and remotely-piloted vehicles) in support of the service components. The directorate also fields UAVs capable of supporting regional strategic intelligence requirements. National-level reconnaissance units are routinely assigned in direct support of operational-level units in the field and can be expected to operate well forward to provide reconnaissance in depth.

### Organization and Mobilization Directorate

2-68. The chief function of the Organization and Mobilization Directorate is to determine the assets the services need to perform operations as directed by the General Staff or SHC. As a result, the directorate both responds to and contributes to the work of the Operations Directorate. The Organization and Mobilization Directorate must also coordinate with other ministries of the government to assure that their planning will meet wartime needs of the plans that the Operations Directorate develops. Planning for determining requirements to mobilize reserves and militia consumes a large percentage of the Organization and Mobilization Directorate's efforts.

2-69. During wartime, the Organization and Mobilization Directorate identifies and recommends priorities of effort in coordination with civilian ministries. It deals particularly with those ministries concerned with transportation and the production or acquisition of food, fuel, and materiel. Due to the directorate's broad charter, its priorities are in effect the State's priorities. Therefore, if it determines a priority, its decision may have the effect of law.

## SERVICE COMPONENTS

2-70. The Armed Forces generally consist of six services. These include the Army, Navy, Air Force, Strategic Forces, Special-Purpose Forces (SPF) Command, and Internal Security Forces. The Internal Security Forces are subordinate to the Ministry of the Interior in peacetime, but become subordinate to the SHC in time of war. The Armed Forces are not a fully unified defense force, but are able to operate jointly as required. The Army is the dominant partner among the military forces. Some functional forces may be organized as joint forces. For example, intelligence and IW units may operate jointly in support of the Intelligence Directorate. The Armed Forces field some reserve component forces in all services, but most reserve forces are Army forces. Militia forces belong exclusively to the ground component.

### Army

2-71. The Army, comprised of ground forces, is the largest and most influential of the State's Armed Forces. This reflects the State's analysis of its potential enemies and the fact that the Armed Forces cannot hope to compete qualitatively with the air and naval components of a major extraregional power such as the United States. The Army is large and, in comparison to most regional neighbors, is modern as well.

2-72. The composition of the Army reflects the great diversity of the geography in which it operates and the Armed Forces' notions about the value of tempo in offensive operations. Accordingly, the Army includes armor, infantry, and a small number of airborne and special-purpose forces. Equally important, the Armed Forces field a large militia, composed primarily of infantry, that is designed to secure LOCs and defend the State if it decides to conduct adaptive operations. The Army's chief offensive power is its armor, mechanized infantry, and airborne forces. The Army also has its own Special-Purpose Forces (Army SPF) designed to support the ground forces at the operational level and enable the Army to conduct reconnaissance and direct action to the opponent's operational depth.

2-73. The Army has adequate resources to conduct regional offensive operations supported and sustained by relatively modern conventional forces. Armor and mechanized infantry forces are the most capable and the most modern of the Army's forces. Militia forces include many elements common to all infantry forces, but are generally not capable of large offensive operations.

2-74. The Army devotes a large share of its means to fielding and maintaining both rocket and tube artillery to support ground operations. The Army also has some long-range rockets and surface-to-surface missiles (SSMs). Fire support capability includes attack helicopters employed in close support. Attack helicopter units also have a limited deep-attack capability. Army air defense is based on the assumption that every soldier with a shoulder-fired surface-to-air missile (SAM) is an air defense firing unit. The Army is assigned large numbers of shoulder-fired SAMs and will also have mobile air defense units in support.

### Navy

2-75. The State does not possess the industrial base to sustain a blue-water, force-projection navy. It does, however, maintain a regional force-projection navy with a significant access-control capability built on small surface combatants and submarines. The Navy can project naval infantry up to brigade size regionally. While the Navy may have difficulty in warding off attack by major surface combatants or air forces of extraregional powers, it is more than adequate to operate against likely regional opponents.

2-76. The Navy is designed primarily to overwhelm regional enemies by means of small fast-attack craft armed with antiship missiles but is also equipped to challenge adversaries' access to the region. Surface- and ground-based antiship missile units form the bulk of the Navy. The Navy also fields a submarine force designed primarily for access-control operations, but also able to insert naval infantry to conduct raids against critical installations within the region. The submarine force, with significant effort ahead of time, may also insert SPF for reconnaissance or direct action outside the region. This kind of operation assumes high risk and is unlikely without expectation of an important payoff.

2-77. The Navy has also invested in antiship technology including wake-homing mines and antiship mines. As a consequence, the Navy has the capability to challenge the most modern navies at maritime choke points, using a combination of mines, shore- and sea-based antiship missiles, and submarines.

2-78. In the event of blockade, the State is able to sustain naval operations from home ports for a limited time. The Navy can sustain force projection regionally for limited duration. A month or more of combat operations that cannot be directly supported without resupply from home ports can challenge the State's capacity. Submarine operations are the exception to this assessment. The Navy may operate submarines outside the region by arrangements with other governments. This capability, while not assured, cannot be ruled out.

2-79. The Navy has a limited amphibious capability and possesses naval infantry capable of conducting forcible entry against regional opponents. The Navy fields organic Special-Purpose Forces (Naval SPF) that are able to conduct reconnaissance in support of landings or raids against critical targets.

2-80. Although the Navy is a coastal navy, it is able to maintain control of inland waterways. This brown-water capability is generally disposed to defend State waterways, but can conduct limited operations in estuaries or further inland in support of Naval SPF or ground combat operations.

## Air Force

2-81. The Air Force, like the Navy, is fundamentally a supporting arm. The State's assessment of the strategic picture is that it cannot hope to compete with the most modern air forces in size or quality. The Air Force has developed a significant regional capability and adequate air defenses to challenge outside air intervention. Its aircraft include fighters, bombers, tactical transport, tankers, airborne early warning (AEW) aircraft, electronic warfare (EW) aircraft, reconnaissance aircraft, and auxiliaries. The Air Force has a combination of small numbers of state-of-the-art air-superiority aircraft, reconnaissance and EW aircraft, and modern, but less than top-line ground-attack aircraft consistent with State force design and procurement principles.

2-82. Along with its air-superiority aircraft, the Air Force has made important investments in AEW, reconnaissance, and EW aircraft. These aircraft are adequate to dominate the airspace against regional opponents. They may also be sufficient to challenge outside forces that intervene in the region, at least in early stages. They may prove effective in supporting access-control operations during transition or adaptive operations. However, the State assumes that these aircraft would not prove viable over the long term against an extraregional threat. In the event of full-scale intervention by modern air forces, the Air Force would not be able to accomplish more than local air parity and would not be able to sustain even that. Due to its investment in ground-based air defenses, however, the State could still be capable of preventing opponents from attaining air supremacy without a long fight.

2-83. The Air Force's bomber aviation forces include a combination of modern and older systems. Deep operations against extraregional forces are unlikely except against high-value targets that the State estimates to be worth the expenditure of a scarce resource. Bombers and other fixed-wing aircraft can have precision munitions and cluster bomb capability. The Air Force also fields limited quantities of fuel-air explosives and standoff systems including some short-range ground-attack and antiship cruise missiles.

2-84. The Air Force has transport aircraft designed to provide two capabilities: force projection and insertion. Theater transports enable the projection of airborne troops regionally. The Air Force also fields light transport aircraft for insertion of its own SPF or those belonging to other components. None of these aircraft affords a legitimate strategic force-projection capability. The Air Force is able to sustain troops on the ground via an air bridge in the absence of a significant air-to-air threat. Transport helicopters supporting ground forces operations may either come from Army aviation units or be allocated from the Air Force.

2-85. The State's national-level Air Defense Forces are subordinate to the Air Force and reflect the State's force development philosophy. They combine obsolescent and state-of-the-art air defense firing units to support area defense and point defense of high-value assets. The State has invested in point-defense technology that it believes can prove effective in defending against cruise missiles.

2-86. Air Force SPF provide organic reconnaissance support and early warning for air defense sectors. They can conduct raids against enemy air bases and installations. They also support search and rescue of aircrews. Finally, SPF form the core for air base security.

2-87. The Air Force manages the State's use of space-based assets and capabilities. The State's reliance on such means is minimal, with the concentration on satellite communication and navigation. The Armed Forces currently have satellite communications (SATCOMs) at the operational level and are in the process of extending this capability down to the tactical level. The SATCOM capabilities the State now uses are commercially-leased circuits. The State possesses limited indigenous space-based RISTA capabilities. However, it can purchase commercially-available imagery products as needed. The increasing ability to leverage commercial satellites will continue to erode the advantage now enjoyed by the United States and other wealthy nations with space-based platforms.

## Strategic Forces

2-88. The Strategic Forces consist of long-range rocket and missile units. The State can develop or purchase a number of theater ballistic missiles (TBMs), as well as intermediate-range ballistic missiles (IRBMs). This constitutes the bulk of the State's strategic capability.

2-89. The missiles of the Strategic Forces are capable of delivering nuclear, biological, and chemical (NBC) munitions, and the NCA is the ultimate NBC release authority. Therefore, the NCA is likely to retain major elements of the Strategic Forces under its direct control or under the SHC or a theater headquarters in wartime.

2-90. The State considers the Strategic Forces capability, even when delivering conventional munitions, the responsibility of the NCA. Long-range missiles and rockets are political tools, first and foremost. The Strategic Forces could use long-range missiles and rockets to advance State political ends during regional, transition, and adaptive operations. Unable to mount robust air campaigns, the State can use these weapons to mount an equivalent effort.

2-91. In some cases, the SHC or theater commander may allocate some Strategic Forces assets down to operational-level commands. Conventionally-armed rocket and missile units may be assigned directly in support of air, naval, and ground forces.

## Special-Purpose Forces Command

2-92. Together with Strategic Forces, SPF provide the State the ability to attack both regional and extraregional enemies throughout their strategic depth. The State will use SPF and insert them in advance of regional operations and in support of transition and adaptive operations. They are an essential element of the concept of using all means necessary and critical to access-control operations. The SPF Command's assigned units provide a balanced capability, including some tactical transport for use in inserting SPF units.

2-93. The General Staff or SHC normally reserves some units of the SPF Command under its own control for strategic-level missions. It may assign some SPF units to subordinate operational or theater commands, but can still task the allocated units to support strategic missions, if required.

2-94. The SPF Command's forces provide a regional and global strategic capability. In addition to conducting direct action, the SPF Command fields strategic reconnaissance forces with which it is able to support national intelligence requirements. It also has a capability to support terrorist and irregular forces operations in other countries. The State may be able to develop means for SPF to deliver nuclear, biological, or chemical munitions.

2-95. Four of the five other service components also have their own SPF. In contrast to the units of the SPF Command, the Army, Navy, and Air Force SPF are designed for use at the operational level. The Internal Security Forces also have their own SPF units. These service SPF normally remain under the control of their respective services or a joint operational or theater command. However, SPF from any of these service components could become part of joint SPF operations in support of national-level requirements. The SPF Command has the means to control joint SPF operations as required.

2-96. Any SPF units (from the SPF Command or from other service components' SPF) that have reconnaissance or direct action missions supporting strategic-level objectives or intelligence requirements would normally be under the direct control of the SHC or under the control of the SPF Command, which reports directly to the SHC. So would any SPF units specially trained and equipped to deliver NBC munitions in support of strategic objectives. Also, any service SPF units assigned to joint SPF operations would temporarily come under the control of the SPF Command or perhaps the SHC.

2-97. The SPF Command also includes elite commando units, specially trained for missions in enemy territory. For administrative purposes, these units may be grouped under a commando brigade headquarters. However, commandos are employed as battalions, companies, platoons, and squads or as small teams, depending on the type of mission. They generally conduct various types of reconnaissance and combat missions in territory not controlled by State forces. Sometimes, particularly in defensive situations, commandos may be called on to perform regular infantry missions, filling gaps between dispersed regular forces.

### Internal Security Forces

2-98. In wartime, some or all of the internal security forces of the Ministry of the Interior become subordinate to the SHC. Thus, they become the sixth service component of the Armed Forces, under the formal designation Internal Security Forces. (For detail, see the Ministry of the Interior section in this chapter or see FM 7-100.3.)

## ADMINISTRATIVE FORCE STRUCTURE

2-99. The State's Armed Forces have an *administrative force structure* that manages military forces in peacetime. This administrative force structure is the aggregate of various military headquarters, organizations, facilities, and installations designed to man, train, and equip the forces. The administrative force structure also has responsibility for disaster management and support to other State agencies.

2-100. The administrative force structure includes all components of the Armed Forces—not only regular, standing forces (active component), but also reserve and militia forces (reserve component). For administrative purposes, both regular and reserve forces come under the headquarters of their respective service component. Each of the six service components is responsible for manning, equipping, and training of its forces and for organizing them within the administrative force structure. The administrative force structure also includes major logistics facilities and installations (see Chapter 7).

### Major Force Groupings

2-101. In peacetime, forces are commonly grouped into corps, armies, or army groups for administrative purposes. In some cases, forces may be grouped administratively under geographical commands designated as military regions or military districts.[1] Normally, these administrative groupings differ from the Armed Forces' go-to-war (fighting) force structure.

2-102. **Peacetime.** Each service commonly maintains its forces grouped under single-service operational-level commands (such as corps, armies, or army groups). A service might also maintain some separate single-service tactical-level commands (divisions, brigades, or battalions) directly under the control of their service headquarters. For example, the Army component headquarters may retain centralized control of certain elite elements of the ground forces, including airborne units and Army SPF. Major tactical-level commands of the Air Force, Navy, Strategic Forces, and the SPF Command often remain under the direct control of their respective service component headquarters. This permits flexibility in the employment of these relatively scarce assets in response to national-level requirements. In peacetime, the internal security forces are under the administrative control of the Ministry of the Interior. (See Figure 2-8.)

2-103. If the General Staff or SHC elects to create more than one theater headquarters, it may allocate parts of the administrative force structure to each of the theaters, normally along geographic lines. One example would be to divide Air Force assets into theater air armies. Another would be to assign units from the SPF Command to each theater, according to theater requirements. During peacetime, however, a separate theater headquarters would exist for planning purposes only and would not have any forces actually subordinated to it.

[1] A military district may or may not coincide with a political district within the State government.

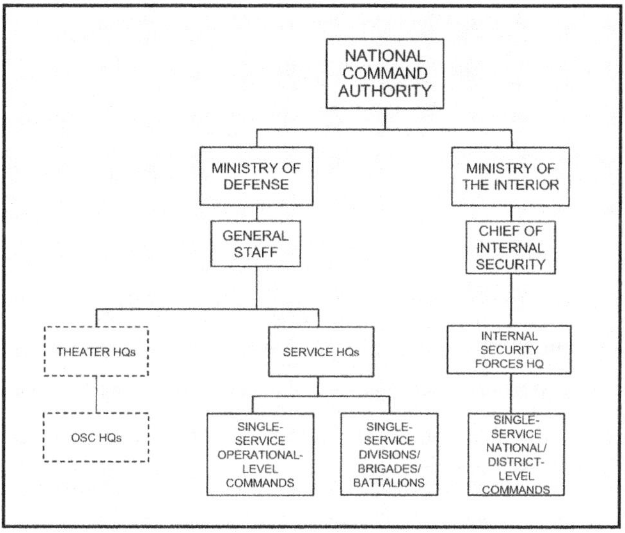


**Figure 2-8. Peacetime Administrative Force Structure**

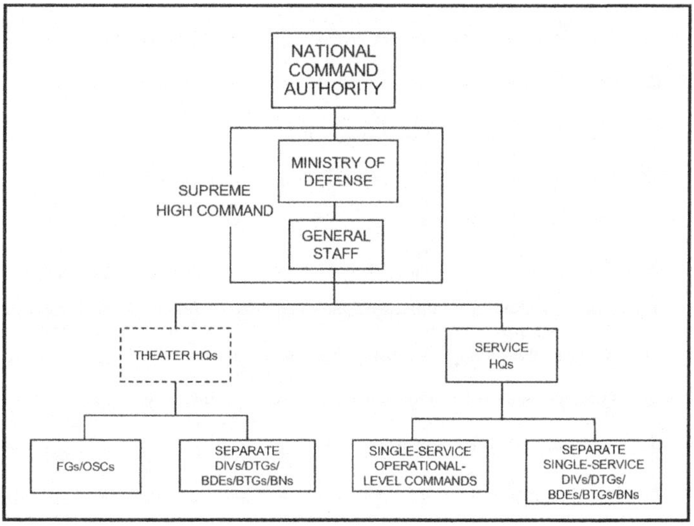


**Figure 2-9. Wartime Fighting Force Structure**

2-104. **Wartime.** In wartime, most major administrative commands continue to exist under their respective service headquarters, but their normal role is to serve as force providers during the creation of operational-level fighting commands, such as field groups (FGs) or operational-strategic commands (OSCs). OSC headquarters may exist in peacetime, for planning purposes, but would not yet have any forces actually subordinate to them. The same would be true of any theater headquarters planned to manage multiple OSCs. FGs, on the other hand, are not normally standing headquarters, but may be organized during full mobilization for war. Some single-service tactical commands may remain under their respective service headquarters or come under the direct control of the SHC or a separate theater headquarters. Some divisions and brigades would be task organized into division tactical groups (DTGs) and brigade tactical groups (BTGs). (See Figure 2-9.)

2-105. After transferring control of its major fighting forces to one or more task-organized fighting commands, an administrative headquarters, facility, or installation continues to provide depot- and area support-level administrative, supply, and maintenance functions. A geographically-based administrative command also provides a framework for the continuing mobilization of reserves to complement or supplement regular forces.

2-106. In rare cases, an administrative command could function as a fighting command. This could occur, for instance, when a particular administrative command happens to have just the right combination of forces for executing a particular strategic campaign plan. Another case would be in times of total mobilization, when an administrative command has already given up part of its forces to a fighting command and then is called upon to form a fighting command with whatever forces remain under the original administrative headquarters.

### Reserves and Militia

2-107. Although all six services field some reserve forces, most of the reserve forces are Army forces. All militia forces belong to the Army component. Overall planning for mobilization of reserves and militia is the responsibility of the Organization and Mobilization Directorate of the General Staff. Each service component headquarters would have a similar directorate responsible for mobilization of forces within that service. Major geographical commands (and other administrative commands at the operational level and higher) serve as a framework for mobilization of reserve and militia forces.

2-108. During mobilization, some reserve personnel serve as individual replacements for combat losses in active units, and some fill positions (including professional and technical specialists) that were left vacant in peacetime in deference to requirements of the civilian sector. However, reservists also man reserve units that are mobilized as units to replace other units that have become combat-ineffective or to provide additional units necessary for large, sustained operations.

2-109. Like active force units, most mobilized reserve and militia units do not necessarily go to war under the same administrative headquarters that controlled them in peacetime. Rather, they typically become part of a task-organized operational- or tactical-level fighting command tailored for a particular mission. In most cases, the mobilized reserve units would be integrated with regular military units in such a fighting command. In rare cases, however, a reserve command at division level or higher might become a fighting command or serve as the basis for forming a fighting command based partially or entirely on reserve forces.

# Chapter 3

# Strategy for Total War

The State's strategic concepts stem from its comparative strength and perceived threats, both internally and regionally, but also take into account the realties of the global strategic environment. Chief among these realities is that, while the State may seek to exploit regional opportunities, intervention by extraregional powers is possible. Therefore, the State assumes that it must be prepared to deal with not only internal and regional opponents, but also these outside forces.

This chapter first explains the overall concepts and principles involved in the State's national security strategy. Then it describes the four basic types of strategic-level courses of action the State leadership uses to execute its strategy for total war in a strategic campaign. Finally, this chapter focuses on the course of action called "strategic operations." For further detail on the other three courses of action, see Chapters 4 through 6.

## NATIONAL SECURITY STRATEGY

3-1. *National security strategy* is the State's vision for itself as a nation and the underlying rationale for building and employing its instruments of national power. It outlines how the State plans to use its diplomatic-political, informational, economic, and military instruments of power to achieve its strategic goals. Among these goals are the State's desires to build and maintain internal stability and to set the conditions for expanding its influence in its region and eventually changing its position within the global community. Formulating national security strategy is a continuous process based on—

- National security interests.
- Threat perceptions.
- Political considerations.
- Foreign policy.
- Economic and military strengths and weaknesses.
- Resources and geography.
- History, religion, and culture.
- Technology.

3-2. Despite the term *security*, the State's national security strategy defines not just what the State wants to protect or defend, but what it wants to achieve. The State's primary strategic goal is continually expanding its influence in the region. This goal is a long-term aim for the State. To the State, it does not matter how long it takes to expand its influence throughout the area, but just that it continually makes expansionary progress. The State is willing to spend the amount of time necessary to achieve its goal. Thus, its national security strategy is designed to allow the State to reach this primary goal.

3-3. The basic national security strategy focuses on the State's expansionary goals in the region. Thus, the State seeks to develop and maintain diplomatic-political, informational, economic, and military power sufficient to dominate the region. However, the State assumes the distinct possibility of extraregional intervention to thwart its regional aspirations. Therefore, to the extent possible, it also invests in technology and establishes forces with that possibility in mind. It builds a doctrine and strategy for employing all these capabilities against all possible threats. Thus, there is no need to improvise if an extraregional force does intervene.

## INTEGRATED PLANNING

3-4. The State's leadership understands the benefits of deliberate planning, perseverance, and willingness to spend the time and other resources necessary to reach the State's desired end state. Methodical planning by the government leaders at the national level results in a well-orchestrated, coordinated State national security strategy for the conduct of total war. This strategy guides the State toward its strategic goals by combining all the State's elements of power in a highly effective and efficient manner.

3-5. The Minister of National Security, through his Strategic Integration Department (SID), integrates all the instruments of national power into one cohesive national security strategy and orchestrates a unified strategic campaign. The State's leadership works well together because its members all share the same ethical, cultural, and religious backgrounds. Leaders from both the ruling class and the diplomatic-political, informational, economic, and military instruments of power participate in strategic planning sessions. They work in consonance to develop and implement a consolidated plan that outlines the State's strategy for total war.

## SYSTEMS WARFARE

3-6. The State defines a *system* as a set of different elements so connected or related as to perform a unique function not performable by the elements or components alone. The essential ingredients of a system include the components, the synergy among components and other systems, and some type of functional boundary separating it from other systems. Therefore, a "system of systems" is a set of different systems so connected or related as to produce results unachievable by the individual systems alone. The State views the operational environment, the battlefield, its own instruments of power, and an opponent's instruments of power as a collection of complex, dynamic, and integrated systems composed of subsystems and components.

3-7. Systems warfare serves as a conceptual and analytical tool to assist in the planning, preparation, and execution of warfare. With the systems approach, the intent is to identify critical system components and attack them in a way that will degrade or destroy the use or importance of the overall system.

## Principle

3-8. The primary principle of systems warfare is the identification and isolation of the critical subsystems or components that give the opponent the capability and cohesion to achieve his aims. The focus is on the disaggregation of the system by rendering its subsystems and components ineffective. While the aggregation of these subsystems or components is what makes the overall system work, the interdependence of these subsystems is also a potential vulnerability. Systems warfare has applicability or impact at all three levels of warfare.

## Application at the Strategic Level

3-9. At the strategic level, the instruments of power and their application are the focus of analysis. National power is a system of systems in which the instruments of national power work together to create a synergistic effect. (See Figure 3-1.) Each instrument of power (diplomatic-political, informational, economic, and military) is also a collection of complex and interrelated subsystems on its own part. The State clearly understands how to analyze and locate the critical components of its own instruments of power and those of its opponents.

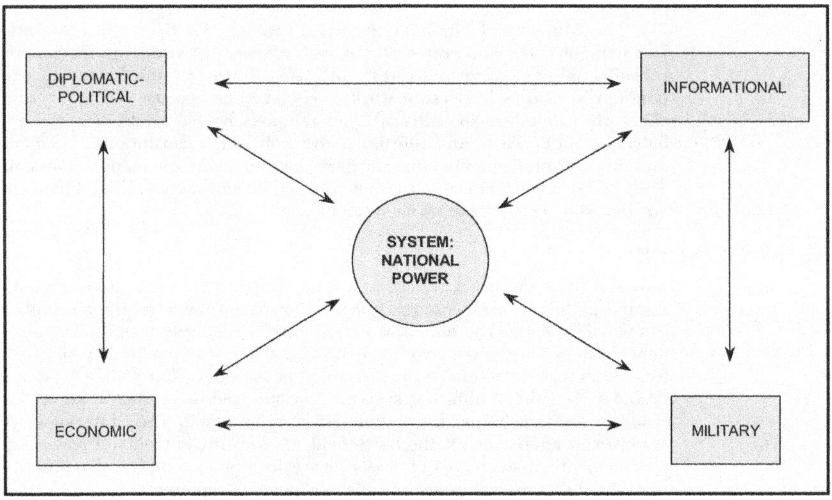


**Figure 3-1. Systems Warfare at Strategic Level**

3-10. Thus, the State can target the systems and subsystems that make up the opponent's instruments of power. At the strategic level, the State can use its own instruments of power to counter or attack an opponent's instruments of power. The primary purpose is to subdue, control, or change the opponent's behavior. If the opponent's strength lies in his military power, the State can attack the other instruments of power as a means of disaggregating or disrupting the system and rendering the overall system ineffective—without necessarily having to defeat the opponent militarily.

3-11. Conversely, the State identifies the critical components of its own instruments of national power and will aggressively aim to protect them from such attack or vulnerabilities. It realizes that the continued functioning of this system of systems as an aggregate is the key to its survival and its aspirations.

### Application at the Operational and Tactical Levels

3-12. When it comes to military confrontation at the operational and tactical levels, systems warfare focuses on disaggregating the enemy's combat system. (See Figure 3-2.) A *combat system* is defined as the combination of weapons systems; organizations; logistics; command and control ($C^2$); and reconnaissance, intelligence, surveillance, and target acquisition (RISTA) that are integrated and regulated by interaction or interdependence to achieve a military function.

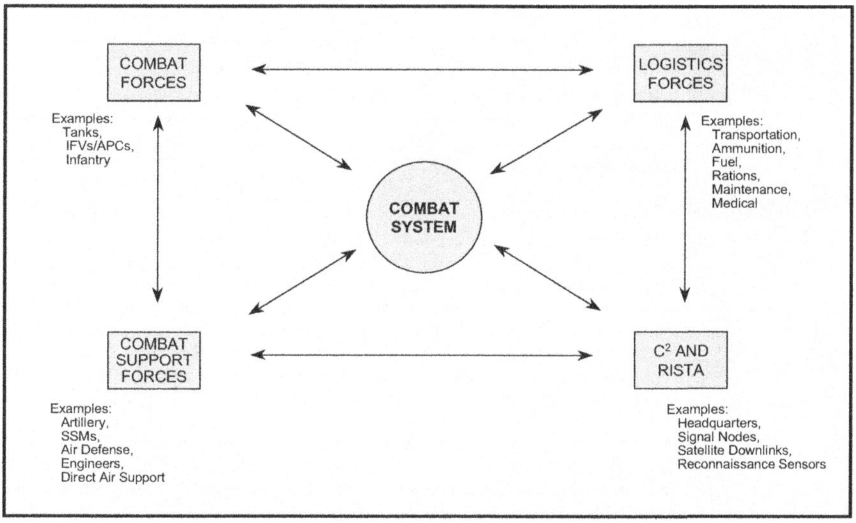


**Figure 3-2. Systems Warfare at Operational and Tactical Levels**

## TOTAL WAR

3-13. The State differentiates between war and armed conflict; war is the more comprehensive of the two. The State's concept of war is "total war," involving the entire country and all its instruments of national power, not just the military. It affects all aspects of life and society. Management of the war effort resides with political leadership. Armed conflict is the aggregate of military actions conducted to attain both military and political strategic goals. It consists primarily of combat and related activities and falls under the management of military leaders.

## ALL MEANS NECESSARY

3-14. Closely related to the concept of "total war" is the concept of "all means necessary." In its broadest sense, the State's approach to war calls for the integrated application of diplomatic-political, informational, economic, and military factors to achieve victory. The State is capable of fielding and maneuvering large formations and recognizes that the employment of those forces in mass has its applications. Nevertheless, the State does not accept that war must be attrition-based. It believes that, through the adaptive use of all available capabilities, it can create opportunities that, properly leveraged, can defeat even a technologically superior opponent. This concept of "all means necessary" simply means that, when threatened, the State will bring to bear all its resources, without exception, to achieve a favorable result.

3-15. Against any potential opponent—internal, regional, or extraregional—the State will deal with the threat(s) not just with the Armed Forces, but with all the elements of power at its disposal. The State will use all its elements of power achieve its strategic goals during peace and war.

3-16. In wartime, State strategy for total war is how the State plans to manage the application and preservation of all its instruments of national power. Thus, it is more than just military strategy (strategy as a higher level of military action than operational art and tactics).

## DISPROPORTIONATE INTERESTS

3-17. When engaged in a conflict within its own country or its own region, the State has more vital interests in the conflict and its outcome than does an extraregional adversary. The goals of the extraregional power for intervening may be political or economic, while the State's goal is expansion and survival. What the extraregional power views as at most a major regional conflict, the State views as total war. The extraregional power may seek to limit the conflict, but the State is fully committed. The State will wage total war, while the extraregional enemy is more than likely waging a limited war. It will commit all means necessary, for as long as necessary, to achieve its strategic goals, while the extraregional power is often tied to a shorter timetable for its commitment. The State will try to force the enemy to fight to get into the region; hitting his aerial and sea ports of debarkation (APODs and SPODs) also has political value.

3-18. Victory may go to the State because it wants to win, needs to win, and is willing to sacrifice to win. Victory need not equate to defeating the extraregional enemy; it is sufficient to just get him to leave the region. In fact, the State may be able to achieve its strategic goals by allowing the enemy to claim victory and depart. In most cases, however, the State will have to try to convince the extraregional power's public and decision makers that the goal of continued intervention is not worth the cost to them. While the extraregional power typically has an aversion to heavy casualties, the State has a much higher threshold for accepting losses, as long as it can preserve its instruments of power. If the State has succeeded in preserving those instruments of power (diplomatic-political, informational, military, and economic), it can reassert its power and influence in the region after the extraregional force departs. Achieving a stalemate or causing a major extraregional power to evacuate the region constitutes a great victory, even if the State suffers losses in doing so. Such a victory can bolster the State's international prestige, foster regional dominance, and strengthen internal support for its national security strategy.

## NATIONAL STRATEGIC GOALS

3-19. The National Command Authority (NCA) determines the State's strategic goals. The State's overall goals are to continually expand its influence within its region and eventually change its position within the global community. These are the long-term aims of the State.

3-20. Supporting the overall, long-term, strategic goals, there may be one or more specific goals, each based on a particular threat or opportunity. Examples of specific strategic goals might be—

- Annexation of territory.
- Economic expansion.
- Destruction of an insurgency.
- Protection of a related minority in a neighboring country.
- Acquisition of natural resources located outside the State's boundaries.
- Destruction of external weapons, forces, or facilities that threaten the existence of the State.
- Defense of the State against invasion.
- Preclusion or elimination of outside intervention.

Each of these specific goals contributes to achieving the overall strategic goals.

## FRAMEWORK FOR IMPLEMENTING NATIONAL SECURITY STRATEGY

3-21. In pursuit of its national security strategy, the State is prepared to conduct four basic types of strategic-level courses of action. Each course of action involves the use of all four instruments of national power, but to different degrees and in different ways. The State gives the four types the following names:

- **Strategic operations**—strategic-level course of action that uses all instruments of power in peace and war to achieve the goals of the State's national security strategy by attacking the enemy's strategic centers of gravity. (For more detail, see the main heading on Strategic Operations section later in this chapter.)
- **Regional operations**—strategic-level course of action (including conventional, force-on-force military operations) against opponents the State overmatches, including regional adversaries and internal threats. (See Chapter 4 for more detail.)

- **Transition operations**—strategic-level course of action that bridges the gap between regional and adaptive operations and contains some elements of both, continuing to pursue the State's regional goals while dealing with the development of outside intervention with the potential for overmatching the State. (See Chapter 5 for more detail.)
- **Adaptive operations**—strategic-level course of action to preserve the State's power and apply it in adaptive ways against opponents that overmatch the State. (See Chapter 6 for more detail.)

Although the State refers to them as "operations," each of these courses of action is actually a subcategory of strategy. Each of these types of "operations" is actually the aggregation of the effects of tactical, operational, and strategic military actions, in conjunction with the other three instruments of national power, that contribute to the accomplishment of strategic goals. The type(s) of operations the State employs at a given time will depend on the types of threats and opportunities present and other conditions in the operational environment. Figure 3-3 illustrates the State's basic conceptual framework for how it could apply its various instruments of national power in the implementation of its national security strategy.

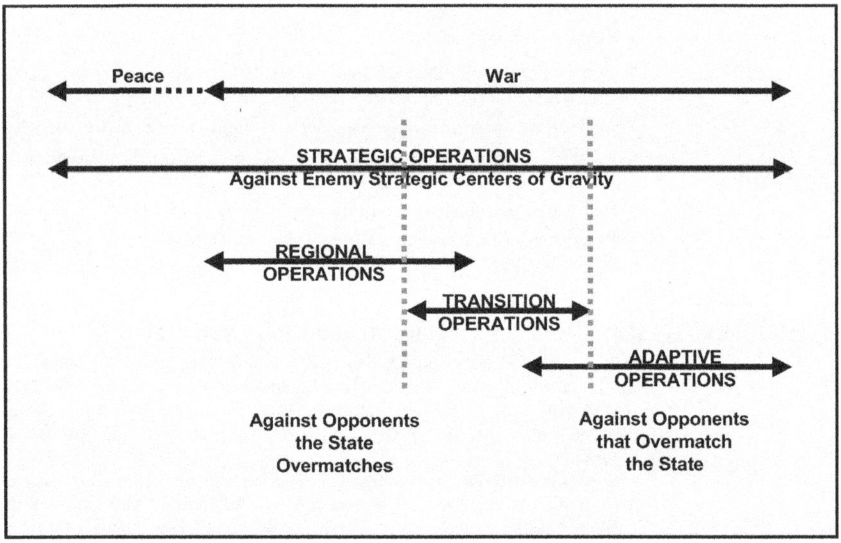


**Figure 3-3. Conceptual Framework for Implementing the State's National Security Strategy**

3-22. Strategic operations are a continuous process not limited to wartime or preparation for war. Once war begins, they continue during regional, transition, and adaptive operations and complement those operations. Each of the latter three types of operations occurs only during war and only under certain conditions. Transition operations can overlap regional and adaptive operations.

3-23. The national security strategy identifies branches, sequels, and contingencies and the role and scope of each type of strategic-level action within these modifications to the basic strategy. Successful execution of these branches and sequels can allow the State to resume regional operations and thus achieve its strategic goals. (See Figure 3-4.)

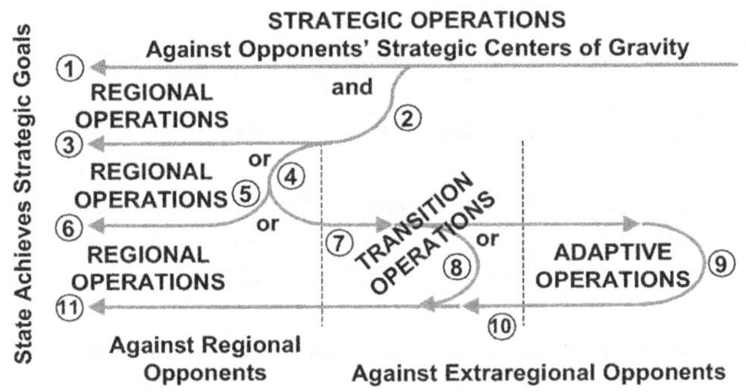


(1) State achieves its strategic goals without combat, through strategic operations that compel other actors to yield to State's will.
(2) State must react to a particular regional threat or seizes an opportunity to change status quo in region.
(3) While strategic operations continue, State achieves its strategic goals through regional operations against weaker neighbors, without extraregional intervention.
(4) Extraregional power threatens to intervene.
(5) State deters extraregional intervention through strategic operations.
(6) State achieves its strategic goals after deterring intervention.
(7) State unable to deter intervention; begins transition operations.
(8) State defeats early-entry forces or causes extraregional forces to withdraw from region; transitions back to regional operations.
(9) Extraregional forces fully deployed and overmatch State forces; State conducts adaptive operations.
(10) Extraregional forces withdraw from region after defeat or stalemate; State begins transition back to regional operations.
(11) State achieves its strategic goals after extraregional intervention.

**Figure 3-4. Examples of Branches and Sequels in National Security Strategy**

3-24. The national security strategy is designed to achieve one or more specific strategic goals within the State's region. Therefore, it typically starts with actions directed at an opponent within the region—an opponent that the State overmatches in conventional military power, as well as other instruments of power.

3-25. The State will attempt to achieve its ends without resorting to armed conflict. Accordingly, strategic operations are not limited to military means and usually do not begin with armed conflict. The State may be able to achieve the desired goal through pressure applied by other-than-military instruments of power, perhaps with the mere threat of using its superior military power against the regional opponent. These actions would fall under the general framework of "strategic operations."

3-26. When nonmilitary means are not sufficient or expedient, the State may resort to armed conflict as a means of creating conditions that lead to the desired end state. However, strategic operations continue even if a particular regional threat or opportunity causes the State to undertake "regional operations" that include military means.

3-27. Prior to initiating armed conflict and throughout the course of armed conflict with its regional opponent, the State continues to conduct strategic operations to preclude intervention by outside players—by other regional neighbors or by an extraregional power that could overmatch the State's forces. However, those operations always include branches and sequels for dealing with the possibility of intervention by an extraregional power.

3-28. When unable to limit the conflict to regional operations, the State is prepared to engage extraregional forces through "transition and adaptive operations." Usually, the State does not shift directly from regional to adaptive operations. The transition is incremental and does not occur at a single, easily identifiable point. If the State perceives intervention is likely, transition operations may begin simultaneously with regional and strategic operations. Transition operations overlap both regional and adaptive operations. Transition operations allow the State to shift to adaptive operations or back to regional operations. At some point, the State either seizes an opportunity to return to regional operations, or it reaches a point where it must complete the shift to adaptive operations. Even after shifting to adaptive operations, the State tries to set conditions for transitioning back to regional operations.

3-29. If an extraregional power were to have significant forces already deployed in the region prior to the outbreak of hostilities, the State would not be able to conduct regional operations using its normal, conventional design without first eliminating those forces. In this case, the State would first use strategic operations—with all means available—to put pressure on the already present extraregional force to withdraw from the region or at least remain neutral in the regional conflict. Barring that, strategic operations could still aim at keeping the extraregional power from committing additional forces to the region and preventing his forces already there from being able to fully exercise their capabilities. If the extraregional force is still able to intervene, the rest of the strategic campaign would have to start with adaptive operations. Eventually, the State would hope to move into transition operations. If it could neutralize or eliminate the extraregional force, it could finally complete the transition to regional operations and thus achieve its strategic goals.

## STRATEGIC CAMPAIGN

3-30. To achieve one or more specific strategic goals, the NCA would develop and implement a specific *national strategic campaign*. Such a campaign is the aggregate of actions of all the State's instruments of power to achieve a specific set of the State's strategic goals against internal, regional, and/or extraregional opponents. There would normally be a diplomatic-political campaign, an information campaign, and an economic campaign, as well as a military campaign. All of these must fit into a single, integrated national strategic campaign.

3-31. The NCA will develop a series of contingency plans for a number of different specific strategic goals that it might want or need to pursue. These contingency plans often serve as the basis for training and preparing the State's forces. These plans would address the allocation of resources to a potential strategic campaign and the actions to be taken by each instrument of national power contributing to such a campaign.

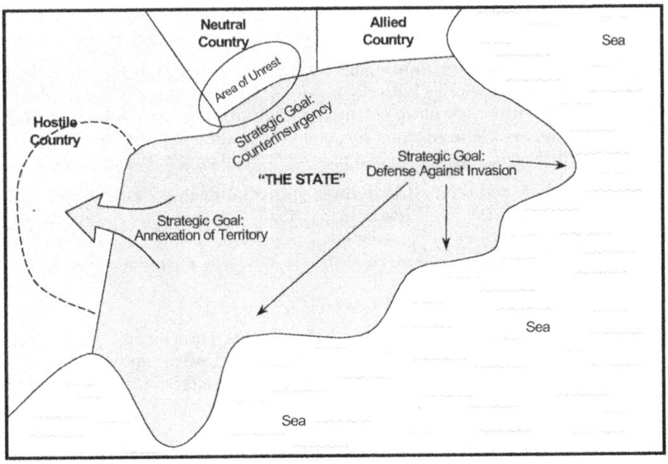


**Figure 3-5. Example of a National Strategic Campaign**

3-32. Aside from training exercises, the NCA would approve only one strategic campaign for implementation at a given time. Nevertheless, the single campaign could include more than one specific strategic goal. For instance, any strategic campaign designed to deal with an insurgency would include contingencies for dealing with reactions from regional neighbors or an extraregional power that could adversely affect the State and its ability to achieve the selected goal. Likewise, any strategic campaign focused on a goal that involves the State's invasion of a regional neighbor would have to take into consideration possible adverse actions by other regional neighbors, the possibility that insurgents might use this opportunity to take action against the State, and the distinct possibility that the original or expanded regional conflict might lead to extraregional intervention. Figure 3-5 shows an example of a single strategic campaign that includes three strategic goals. (The map of the State in this diagram is for illustrative purposes only and does not necessarily reflect the actual size, shape, or physical environment of the State or its neighbors.)

## NATIONAL STRATEGIC CAMPAIGN PLAN

3-33. The purpose of a *national strategic campaign plan* (national SCP) is to integrate all the instruments of national power under a single plan. Even if the State hoped to achieve the goal(s) of the campaign by nonmilitary means, the national campaign plan would leverage the influence of its Armed Forces' strong military presence and provide for the contingency that military force might become necessary.

3-34. The national SCP is the end result of the SID's planning effort. Based on input from all State ministries, this is the plan for integrating the actions of all instruments of power to set conditions favorable for achieving the central goal identified in the national security strategy. The Ministry of Defense (MOD) is only one of several ministries that provide input and are then responsible for carrying out their respective parts of the consolidated national plan.

3-35. In waging a national strategic campaign, the State never employs military power alone. Military power is most effective when applied in combination with diplomatic-political, informational, and economic instruments of power. State ministries responsible for each of the four instruments of power will develop their own campaign plans as part of the unified national SCP.

3-36. A national SCP defines the relationships among all State organizations, military and nonmilitary, for the purposes of executing that SCP. The SCP describes the intended integration, if any, of multinational forces in those instances where the State is acting as part of a coalition.

## MILITARY STRATEGIC CAMPAIGN PLAN

3-37. During peacetime, the Operations Directorate of the General Staff is responsible for developing, staffing, promulgation, and continuing review of the *military strategic campaign plan*. It must ensure that the military plan would end in achieving military conditions that would fit with the conditions created by the diplomatic-political, informational, and economic portions of the national plan that are prepared by other State ministries. Therefore, the Operations Directorate assigns liaison officers to other important government ministries.

3-38. In wartime, the MOD and the General Staff combine to form the Supreme High Command (SHC). The Operations Directorate continues to review the military SCP and modify it or develop new plans based on guidance from the Chief of the General Staff (CGS), who commands the SHC. It generates options and contingency plans for various situations that may arise. Once the CGS approves a particular plan for a particular strategic goal, he issues it to the appropriate operational-level commanders.

3-39. The military SCP directs operational-level military forces, and each command identified in the SCP prepares an operation plan that supports the execution of its role in that SCP. The SCP assigns forces to operational-level commands and designates areas of responsibility for those commands.

3-40. From the General Staff down through the operational and tactical levels, the staff of each military headquarters has an operations element that is responsible for planning. The plan at each level specifies the task organization of forces allocated to that level of command, in order to best accomplish the mission assigned by a higher headquarters. Once the commander at a particular level approves the plan, he issues it to the subordinate commanders who will execute it.

## THEATER

3-41. For the State, a *theater* is a clearly defined geographic area in which the State's Armed Forces plan to conduct or are conducting military operations. However, the term *theater* may have a different meaning for the State than for a major extraregional power. For an extraregional power with global force-projection capability, a theater is any one of several geographic areas of the world where its forces may become involved. For the State, however, the only theater (or theaters) in question would be within the region of the world in which the State is located and is capable of exerting its regionally-centered power. The extraregional power may not define the limits of this specific region in exactly the same way that the State defines it, in terms of its own perceptions and interests. Within its region, the State may plan or conduct a strategic campaign in a single theater or in multiple theaters, depending on the situation.

3-42. The General Staff may create one or more separate theater headquarters even in peacetime, for planning purposes. However, no forces would be subordinated to such a headquarters until the activation of a particular SCP.

3-43. When there is only one theater, as is typical, the theater headquarters may also be the field headquarters of the SHC, and the CGS may also be the theater commander. Even in this case, however, the CGS may choose to focus his attention on national strategic matters and to create a separate theater headquarters, commanded by another general officer, to control operations within the theater. A theater headquarters provides flexible and responsive control of all theater forces.

3-44. When parts of the strategic campaign take place in separated geographical areas and there is more than one major line of operations, the State may employ more than one theater headquarters, each of which could have its own theater campaign plan. In this case, albeit rare, the SHC field headquarters would be a separate element exercising control over the multiple theater headquarters.

3-45. The existence of one or more separate theater headquarters provides flexibility to the State, in that it enables the SHC to focus on the strategic campaign and to sustain the forces in the field. The theater headquarters acts to effectively centralize and integrate General Staff control over theater-wide operations and to exercise command over all forces assigned to a theater in accordance with mission and aim assigned by the SHC.

3-46. For the State, a theater (or theaters) may include only those parts of the region where it wishes to exert its own power and influence. However, the theater(s) could also extend to areas where the State wishes to prevent its regional neighbors or extraregional powers from exerting their power and influence. During war, therefore, the State may extend its definition of a given theater to include staging areas from which other forces can deploy into the theater proper. In the event of threats from extraregional powers, the State may authorize "out-of-theater" operations to neutralize a potential opponent's power projection capabilities. For the State, the concept of *theater(s)* is part of the concept of *total war*, which is not entirely military.

## STRATEGIC OPERATIONS

3-47. The State's "strategic operations" involve the application of any or all of the four instruments of national power at the direction of the *national-level* decision makers in the NCA. They occur throughout the strategic campaign. The nature of strategic operations at any particular time corresponds to the conditions perceived by the NCA. These operations also differ from the other operations of a strategic campaign in that they are not limited to wartime and can transcend the region. (For more detail, see the main heading on Strategic Operations later in this chapter.)

3-48. During regional, transition, or adaptive operations, the State intends to employ all its instruments of power in strategic operations against the enemy's strategic centers of gravity. Even before the outset of regional operations, the State lays plans to prevent outside intervention in the region while defeating its internal or regional enemies or otherwise achieving its strategic goals. If extraregional intervention occurs, the State continues to employ strategic operations while conducting transition and adaptive operations. Strategic operations typically target intangible elements of the enemy's efforts against the State, including his will to fight, public support, and alliances or coalitions.

## REGIONAL OPERATIONS

3-49. The State possesses an overmatch in most, and sometimes all, elements of power against regional opponents. It is able to employ that power in a conventional operational design focused on offensive action. A weaker regional neighbor may not actually represent a threat to the State, but rather an opportunity that the State can exploit. (For more detail on regional operations, see Chapter 4.)

3-50. In such conditions, the State's doctrine, organization, capabilities, and national security strategy allow it to deal with internal and regional threats or opportunities primarily through offensive action. This enables the State to control its own population and dominate neighboring nations, at least at the outset. The State plans these operations well in advance and executes them as rapidly as possible, in order to preclude regional alliances or outside intervention.

3-51. The State designs its military forces and employs an investment strategy that ensures superiority in conventional military power over any of its regional neighbors. Regionally-focused operations typically involve "conventional" patterns of operation. However, the term *conventional* does not mean that the

State will use only conventional forces and conventional weapons in such a conflict, nor does it mean that the State will not use some adaptive approaches.

3-52. During regional operations, the State also relies on its continuing strategic operations to preclude or control outside intervention. It tries to keep foreign perceptions of its actions during a regional conflict below the threshold that will invite in extraregional forces. The State wants to win the regional conflict, but has to be careful how it does so. It works to prevent development of international consensus for intervention and to create doubt among possible participants.

## TRANSITION OPERATIONS

3-53. Transition operations serve as a pivotal point between regional and adaptive operations. The transition may go in either direction. The fact that the State begins transition operations does not necessarily mean that it must complete the transition from regional to adaptive operations (or vice versa). As conditions allow or dictate, the "transition" could end with the State conducting the same type of operations as before the shift to transition operations. (For more detail on transition operations, see Chapter 5.)

3-54. The State conducts transition operations when other regional and/or extraregional forces threaten the State's ability to continue regional operations in a conventional design against the original regional enemy. At the point of shifting to transition operations, the State still has the ability to exert all instruments of national power against an overmatched regional enemy. Indeed, it may have already defeated its original adversary. However, its successful actions in regional operations have prompted either other regional neighbors or an extraregional power to contemplate intervention. The State will use all means necessary to preclude or defeat intervention.

3-55. Although the State would prefer to achieve its strategic goal through regional operations, an SCP has the flexibility to be able to change and adapt if required. Since the State assumes the possibility of extraregional intervention, any SCP will already contain thorough plans for transition operations, as well as adaptive operations, if necessary.

3-56. When an extraregional force starts to deploy into the region, the balance of power begins to shift away from the State. Although the State may not yet be totally overmatched by the enemy force, it faces a threat it will not be able to handle with normal, "conventional" patterns of operation designed for regional conflict. Therefore, the State must begin to adapt its operations to the changing threat.

3-57. While the State and its Armed Forces as a whole are in the condition of transition operations, an operational- or tactical-level commander will still receive a mission statement in plans and orders from higher headquarters stating the purpose of his actions. To accomplish that purpose and mission, he will use as much as he can of the conventional patterns of operation that were available to him during regional operations and as much as he has to of the more adaptive-type approaches dictated by the presence of an extraregional force.

3-58. Even extraregional forces may be vulnerable to "conventional" operations during the time they require to build combat power and create support at home for their intervention. Against an extraregional force that either could not fully deploy or has been successfully separated into isolated elements, the State's Armed Forces may still be able to use some of the more conventional patterns of operation. The State will not shy away from the use of military means against an advanced extraregional opponent so long as the risk is commensurate with potential gains.

3-59. Transition operations serve as a means for the State to retain the initiative and still pursue its overall strategic goal of regional expansion despite its diminishing advantage in the balance of power. From the outset, one part of the set of specific goals for any strategic campaign was the goal to defeat any outside intervention or prevent it from fully materializing. As the State begins transition operations, its immediate goal is preservation of its instruments of power while seeking to set conditions that will allow it to transition back to regional operations. Transition operations feature a mixture of offensive and defensive actions that help the State control the strategic tempo while changing the nature of conflict to something for which the intervening force is unprepared. Transition operations can also buy time for the State's strategic operations to succeed.

3-60. There are two possible outcomes to transition operations. If the extraregional force suffers sufficient losses or for other reasons must withdraw from the region, the State's operations may begin to transition back to regional operations, again becoming primarily offensive. If the extraregional force is not compelled to withdraw and continues to build up power in the region, the State's transition operations may begin to gravitate in the other direction, toward adaptive operations.

## ADAPTIVE OPERATIONS

3-61. Generally, the State conducts adaptive operations as a consequence of intervention from outside the region. Once an extraregional force intervenes with sufficient power to overmatch the State, the full conventional design used in regionally-focused operations is no longer sufficient to deal with this threat. The State has developed its doctrine, organization, capabilities, and strategy with an eye toward dealing with both regional and extraregional opponents. It has already planned how it will adapt to this new and changing threat and has included this adaptability in its doctrine. (For more detail on adaptive operations, see Chapter 6.)

3-62. Having begun the process of adaptation during transition operations, the State must continue to adapt its operations in order to address the growing threat. Thus, it completes the shift from regional operations to adaptive operations. Adaptive operations help the State to mitigate its disadvantages and exploit its advantages, as compared to the extraregional threat. The State must also continue to deal with internal and regional threats as it prepares to deal with the extraregional threat.

3-63. The longer the State can delay effective extraregional response to the crisis in the region, the greater its chances for success. Failing to limit or interrupt access to the region, the State will attempt to degrade further enemy force projection, hold initial gains, and extend the conflict, while preserving its own military capability and other instruments of national power.

3-64. The State believes that adaptive operations can lead to several possible outcomes. If the results do not completely resolve the conflict in the State's favor, they may at least allow the State to return to regional operations. Even a stalemate may be a victory for the State, as long as it preserves enough of its instruments of power to preserve the regime and lives to fight another day.

3-65. When an extraregional power intervenes with sufficient force to overmatch the State's, the State and its Armed Forces have to adapt their patterns of operation. The State still has the same forces and technology that were available to it for regional operations, but must use them in creative and adaptive ways. It has already thought through how it will adapt to this new or changing threat in general terms. (See Principles of Operation versus an Extraregional Power below.) It has already developed appropriate branches and sequels to its basic SCP and does not have to rely on improvisation. During the course of combat, it will make further adaptations, based on experience and opportunity.

3-66. When the State shifts to adaptive operations, these are often sanctuary-based and more defensive in nature than were regional or transition operations. When overmatched in conventional power, the State seeks to preserve its own power and apply it in adaptive ways. It expects its commanders to seize opportunity, tailor organizations to the mission, and make creative use of existing capabilities even more than they did in regional and transition operations.

3-67. As part of the State's strategy for total war, these adaptive operations attempt to attack the intervening force throughout its depth and to destroy its will and ability to fight. The intent is to delay, disrupt, wear down, and ultimately defeat the intervening force by the application of all means necessary. These operations will use various combinations of military means, including offense and defense, as well as paramilitary and nonmilitary means. They can also use various combinations of weapons, possibly including weapons of mass destruction (WMD).

## PRINCIPLES OF OPERATION VERSUS AN EXTRAREGIONAL POWER

3-68. The State assumes the distinct possibility of intervention by a major extraregional power in any regional conflict. It views the United States as the most advanced extraregional force it might have to face. Like many other countries, the State has studied U.S. military forces and their operations and is pursuing lessons learned based on its assessments and perceptions. The State is therefore using the United States as its baseline for planning adaptive approaches for dealing with the strengths and weaknesses of an extraregional force. It believes that preparing to deal with intervention by U.S. forces will enable it to deal effectively with those of any other extraregional power. Consequently, it has devised the following principles for applying its various instruments of diplomatic-political, informational, economic, and military power against this type of threat.

### Control Access into Region

3-69. Extraregional enemies capable of achieving overmatch against the State must first enter the region using power-projection capabilities. Therefore, the State's force design and investment strategy is focused on access control—to selectively deny, delay, and disrupt entry of extraregional forces into the region and to force them to keep their operating bases beyond continuous operational reach. This is the easiest manner of preventing the accumulation of enemy combat power in the region and thus defeating a technologically superior enemy.

3-70. Access-control operations are continuous throughout the strategic campaign and can reach beyond the theater as defined by the State's NCA. They can begin even before the extraregional power declares its intent to come into the region, and continue regardless of whether the State is conducting regional, transition, or adaptive operations. Access-control operations come in three basic forms: strategic preclusion, operational exclusion, and access limitation.

3-71. **Strategic Preclusion.** *Strategic preclusion* seeks to completely deter extraregional involvement or severely limit its scope and intensity. The State would attempt to achieve strategic preclusion in order to reduce the influence of the extraregional power or to improve its own regional or international standing. It would employ all its instruments of power to preclude direct involvement by the extraregional power. Actions can take many forms and often contain several lines of operation working simultaneously.

3-72.The primary target of strategic preclusion is the extraregional power's national will. First, the State would conduct diplomatic and perception management activities aimed at influencing regional, transnational, and world opinion. This could either break apart ad hoc coalitions or allow the State to establish a coalition of its own or at least gain sympathy. For example, the State might use a disinformation campaign to discredit the legitimacy of diplomatic or economic sanctions imposed upon it. The extraregional power's economy and military would be secondary targets, with both practical and symbolic goals. This might include using global markets and international financial systems to disrupt the economy of the extraregional power, or conducting physical and information attacks against critical economic centers. Similarly, the military could be attacked indirectly by disrupting its power projection, mobilization, and training capacity. Preclusive actions are likely to increase in intensity and scope as the extraregional power moves closer to military action. If strategic preclusion fails, the State will turn to operational methods that attempt to limit the scope of extraregional involvement or cause it to terminate quickly.

3-73. **Operational Exclusion.** *Operational exclusion* seeks to selectively deny an extraregional force the use of or access to forward bases of operation within the region or even outside the theater defined by the NCA. For example, through diplomacy, economic or political connections, information campaigns, and/or hostile actions, the State might seek to deny the enemy the use of bases in other foreign nations. It might also attack population and economic centers for the intimidation effect, using long-range missiles, WMD, or Special-Purpose Forces (SPF).

3-74. Forces originating in the enemy's homeland must negotiate long and difficult air or surface lines of communication (LOCs) merely to reach the region. Therefore, the State will use any means at its disposal to also strike the enemy forces along routes to the region, at transfer points en route, at aerial and sea ports of embarkation (APOEs and SPOEs), and even at their home stations. These are fragile and convenient targets in support of transition and adaptive operations.

3-75. **Access Limitation.** *Access limitation* seeks to affect an extraregional enemy's ability to introduce forces into the theater. Access-control operations do not necessarily have to deny the enemy access entirely. A more realistic goal is to limit or interrupt access into the theater in such a way that the State's forces are capable of dealing with them. By controlling the amount of force or limiting the options for force introduction, the State can create conditions that place its conventional capabilities on a par with those of an extraregional force. Capability is measured in terms of what the enemy can bring to bear in the theater, rather than what the enemy possesses.

3-76. The State's goal is to limit the enemy's accumulation of applicable combat power to a level and to locations that do not threaten the accomplishment of a strategic campaign. This may occur through many methods. For example, the State may be able to limit or interrupt the enemy's deployment through actions against his APODs and SPODs in the region. Hitting such targets also has political and psychological value. The State will try to disrupt and isolate enemy forces that are in the region or coming into it, so that it can destroy them piecemeal. It might exploit and manipulate international media to paint foreign intervention in a poor light, decrease international resolve, and affect the force mix and rules of engagement (ROE) of the deploying extraregional forces.

## Employ Operational Shielding

3-77. The State will use any means necessary to protect key elements of its combat power from destruction by an extraregional force—particularly by air and missile forces. This protection may come from use of any or all of the following:

- Complex terrain.
- Noncombatants.
- Risk of unacceptable collateral damage.
- Countermeasure systems.
- Dispersion.
- Fortifications.
- Information warfare (IW).

3-78. Operational shielding generally cannot protect the entire force for an extended time period. Rather, the State will seek to protect selected elements of its forces for enough time to gain the freedom of action necessary to prosecute important elements of a strategic campaign.

## Control Tempo

3-79. The State initially employs rapid tempo to conclude regional operations before an extraregional force can be introduced. It will also use rapid tempo to set conditions for access-control operations before the extraregional force can establish a foothold in the region. Once it has done that, it needs to be able to control the tempo—to ratchet it up or down, as is advantageous to its own operational or tactical plans.

3-80. During the initial phases of an extraregional enemy's entry into the region, the State's Armed Forces may employ a high operational tempo to take advantage of the weaknesses inherent in enemy power projection (lightly equipped forces are usually the first to enter the region). This may take the form of attack by the Armed Forces against enemy early-entry forces, linked with diplomatic, economic, and informational efforts to terminate the conflict quickly before main enemy forces can be brought to bear. Thus, the State may be able to force the enemy to conventional closure, rather than needing to conduct adaptive operations later, when overmatched by the enemy.

3-81. An extraregional enemy normally tries to slow the tempo while it is deploying into the region and to speed it up again once it has built up overwhelming force superiority. The State's forces will try to increase the tempo when the enemy wants to slow it and to slow the tempo at the time when the enemy wants to accelerate it.

3-82. By their nature, offensive operations tend to control time or tempo and defensive operations tend to determine space or location. Through a combination of defensive and offensive actions, the State's adaptive operations seek to control both location and tempo.

3-83. If the State cannot end the conflict quickly, it may take steps to slow the tempo and prolong the conflict, taking advantage of enemy lack of commitment over time. The preferred Armed Forces tactics during this period would be the ambush and raid as a means of avoiding decisive combat with superior forces. These activities may not be linked to maneuver or ground objectives, but intended instead to inflict mass casualties or destroy flagship systems, both of which will reduce the enemy's will to continue the fight.

## Cause Politically Unacceptable Casualties

3-84. The State will try to inflict highly visible and embarrassing losses on enemy forces to weaken the enemy's domestic resolve and national will to sustain the deployment or conflict. Modern wealthy nations have shown an apparent lack of commitment over time, and sensitivity to domestic and world opinion in relation to conflict and seemingly needless casualties. The State believes it can have a comparative advantage against superior forces because of the collective psyche and will of the people of the State to endure hardship or casualties, while the enemy may not be willing to do the same.

3-85. The State also has the advantage of disproportionate interests: the extraregional force may have limited objectives and only casual interest in the conflict, while the State approaches it from the perspective of total war and the threat to its aspirations or even its national survival. The State is willing to commit all means necessary, for as long as necessary, to achieve its strategic goals. Compared to the extraregional enemy, the State stands more willing to absorb higher military and civilian casualties in order to achieve victory. It will try to influence public opinion in the enemy's homeland to the effect that the goal of intervention is not worth the cost.

3-86. Battlefield victory does not always go to the best-trained, best-equipped and most technologically advanced force. National will encompasses a unification of values, morals, and effort among the population, the government leadership, and their forces. Through this unification, all parties are willing to individually sacrifice for the achievement of the unified goal. The interaction of military actions and political judgements, conditioned by national will, serves to further define and limit the achievable objectives of a conflict for all parties involved, and to determine its duration and conditions of termination.

## Neutralize Technological Overmatch

3-87. Against an extraregional force, the State's Armed Forces will forego massed formations, patterned echelonment, and linear operations that would present easy targets for such an enemy. It will hide and disperse its forces in areas of sanctuary that limit the enemy's ability to apply his full range of technological capabilities. However, the State can rapidly mass forces and fires from those dispersed locations for decisive combat at the time and place of its own choosing.

3-88. The State will attempt to use the physical environment and natural conditions to neutralize or offset the technological advantages of a modern extraregional force. It trains its forces to operate in adverse weather, limited visibility, rugged terrain, and urban environments that shield them from the effects of the enemy's high-technology weapons and deny the enemy the full benefits of his advanced $C^2$ and RISTA systems.

3-89. The State can also use the enemy's robust array of RISTA systems against him. His large numbers of sensors can overwhelm his units' ability to receive, process, and analyze raw intelligence data and to provide timely and accurate intelligence analysis. The State can add to this saturation problem by using deception to flood enemy sensors with masses of conflicting information. Conflicting data from different sensors at different levels (such as satellite imagery conflicting with data from unmanned aerial vehicles) can confuse the enemy and degrade his situational awareness.

3-90. The destruction of high-visibility or unique systems employed by enemy forces offers exponential value in terms of increasing the relative combat power of the State's Armed Forces and maximizes effects in the information and psychological arenas. These actions are not always linked to military objectives. High-visibility systems that could be identified for destruction could include stealth aircraft, attack helicopters, counterbattery artillery radars, aerial surveillance platforms, or rocket launcher systems. Losses among these premier systems may undermine enemy morale, degrade operational capability, and inhibit employment of these weapon systems. The destruction or degradation of these systems may be achieved through the use of persistent chemical strikes, unconventional forces, or conventional raids and ambushes.

3-91. When conducting actions against a superior foe, State forces must seek to operate on the margins of enemy technology and maneuver during periods of reduced exposure, those periods identified by a detailed study of enemy capabilities. If available, precision munitions can degrade or eliminate high-technology weaponry, and sophisticated camouflage, deception, decoy, or mockup systems can degrade the effects of enemy systems. Also, State forces can employ low-cost GPS jammers to disrupt enemy precision munitions targeting, sensor-to-shooter links, and navigation.

3-92. Modern militaries rely upon information and information systems to plan and conduct operations. For this reason, the State will conduct extensive information attacks and other offensive IW actions. It could physically attack enemy systems and critical $C^2$ nodes, or conduct "soft" attacks by utilizing computer viruses or denial-of-service activities. Attacks should target enemy military and civilian decision makers and key information nodes such as information network switching centers, transportation centers, and aerial platform ground stations. Conversely, State information systems and procedures should be designed to deny information to the enemy and protect friendly forces and systems with a well-developed defensive IW plan.

3-93. The State may have access to commercial products to support precision targeting and intelligence analysis. This proliferation of advanced technologies permits organizations to achieve a situational awareness of enemy deployments and force dispositions formerly reserved for selected militaries. Intelligence can also be obtained through greater use of human intelligence (HUMINT) assets that gain intelligence through civilians or local workers contracted by the enemy for base operation purposes. Similarly, technologies such as cellular telephones are becoming more reliable and inexpensive. It is becoming harder to discriminate between use of such systems by civilian and military actors. Therefore, they could act as a primary communications system or a redundant measure of communication, and there is little the enemy can do to prevent the use of these assets.

### Change the Nature of Conflict

3-94. The State will try to change the nature of conflict to exploit the differences between friendly and enemy capabilities. Following an initial period of regionally-focused conventional operations and utilizing the opportunity afforded by phased enemy deployment, the State will change its operations to focus on preserving combat power and exploiting enemy ROE. This change of operations will present the fewest targets possible to the rapidly growing combat power of the enemy. It is possible that power-projection forces, optimized for a certain type of maneuver warfare, would be ill suited to continue operations (for example, a heavy-based projection force confronted with combat in complex terrain).

3-95. Against early-entry forces, the State may still be able to use the design it employed in previous operations against regional opponents. However, as the extraregional force builds up to the point where it threatens to overmatch State forces, the State is prepared to disperse its forces and employ them in patternless operations that present a battlefield that is difficult for the enemy to analyze and predict.

3-96. The State may hide and disperse its forces in areas of sanctuary. The sanctuary may be physical, often located in urban areas or other complex terrain that limits or degrades the capabilities of enemy systems. However, the State may also use moral sanctuary by placing its forces in areas shielded by civilians or close to sites that are culturally, politically, economically, or ecologically sensitive. The State's forces will defend in sanctuaries, when necessary. However, elements of those forces will move out of sanctuaries and attack when they can create a window of opportunity or when opportunity is presented by physical or natural conditions that limit or degrade the enemy's systems.

3-97. The State's forces do not avoid contact; rather, they often seek contact, but on their own terms. Their preferred tactics under these conditions would be the ambush and raid as a means of avoiding decisive combat with superior forces. They will also try to mass fires from dispersed locations to destroy key enemy systems and formations. However, when an opportunity presents itself, the State can rapidly mass forces and adopt more patterned operations for decisive combat.

3-98. The strengths and weaknesses of an adversary may require other adjustments. The State will capitalize on interoperability issues among the enemy forces and their allies by conducting rapid actions before the enemy can respond with overwhelming force. If the State borders another country with a sympathetic population, it can use border areas to provide refuge or a base of attack for insurgent forces. Also, the State can use terror tactics against enemy civilians or soldiers not directly connected to the intervention as a device to change the fundamental nature of the conflict.

3-99. The State may have different criteria for victory than the extraregional force—a stalemate may be good enough. Similarly, its definition of victory may not require a convincing military performance. For example, it may call for inflicting numerous casualties to the enemy. The State's perception of victory may equate to national survival. So the nature of the conflict may be perceived differently in the eyes of the State versus those of the enemy.

### Allow No Sanctuary

3-100. Along with dispersion, decoys, and deception, the State uses urban areas and other complex terrain as sanctuary from the effects of enemy forces. Meanwhile, its intent is to deny enemy forces the use of such terrain. This forces the enemy to operate in areas where the State's long-range fires and strikes can be more effective.

3-101. The State seeks to deny enemy forces safe haven during every phase of a deployment and as long as they are in the region. The resultant drain on manpower and resources to provide adequate force-protection measures can reduce the enemy's strategic, operational, and tactical means to conduct war and erode his national will to sustain conflict. Terror tactics are one of the effective means to deny sanctuary to enemy forces. Terrorism has a purpose that goes well beyond the act itself; the goal is to generate fear. For the State, these acts are part of the concept of total war. State-sponsored or independent terrorists can attack the enemy anywhere and everywhere. The State's SPF can also use terror tactics and are well equipped, armed, and motivated for such missions.

3-102. The State is prepared to attack enemy forces anywhere on the battlefield, at overseas bases, at home stations, and even in military communities. It will attack his airfields, seaports, transportation infrastructures, and LOCs. These attacks feature coordinated operations by all available forces, using not just terror tactics, but possibly long-range missiles and WMD. Targets include not only enemy military forces, but also contractors and private firms involved in transporting troops and materiel into the region. The goal is to present the enemy with a nonlinear, simultaneous battlefield. Striking such targets will not only deny the enemy sanctuary, but also weaken his national will, particularly if the State can strike targets in the enemy's homeland.

## STRATEGIC OPERATIONS

3-103. What the State calls "strategic operations" is actually a universal strategic course of action the State would use to deal with all situations—in peacetime and war, against all kinds of opponents, potential opponents, or neutral parties. Strategic operations involve the State's use of any and all of its instruments of national power to affect the enemy's strategic centers of gravity. In relation to an extraregional power, the first aim of strategic operations is to preclude such a power from intervening in the State's region. If preclusion is not possible, the aim becomes that of getting the extraregional force to leave before it can achieve the goals of its intervention.

### MEANS

3-104. Strategic operations apply all four instruments of power, in varying combinations depending on the conditions. In most cases, the diplomatic-political, informational, and economic means tend to dominate. During strategic operations, military means are most often used to complement those other instruments of power to achieve State goals. For example, the military means are likely to be used against key political or economic centers or tangible targets whose destruction affects intangible centers of gravity, rather than against military targets for purely military objectives.

3-105. Against such targets, the State will employ all means available: diplomatic initiatives, IW, economic pressure, terrorist attacks, State-sponsored insurgency, direct action by SPF, long-range precision fires, and even WMD against selected targets. These efforts often place noncombatants at risk and aim to apply diplomatic-political, economic, and psychological pressure by allowing the enemy no sanctuary.

3-106. The use of diplomatic-political or economic means or pressure is always orchestrated at the national level, as is strategic IW. Even with the military instrument of power, actions considered part of strategic operations require a conscious, calculated decision and direction or authorization by the NCA, which is not entirely military in its makeup.

### TARGETS

3-107. Strategic operations target the enemy's strategic centers of gravity. They attack the intangible components of the enemy's efforts against the State. They primarily target those elements that can most affect factors such as enemy soldiers' and leaders' confidence, political and diplomatic decisions, public opinion, the interests of private institutions, national will, and the collective will and commitment of alliances and coalitions. National will is not just the will to fight, but also the will to intervene by other than military means.

3-108. It may not be readily apparent to outside parties whether specific actions by the State's various instruments of power are part of strategic operations or part of another strategic-level course of action occurring simultaneously. In fact, one action could conceivably fulfill both purposes. For example, a demoralizing defeat that could affect the enemy's strategic centers of gravity could also be a defeat from an operational or tactical viewpoint. In other cases, a particular action on the battlefield might not make sense from a tactical or operational viewpoint, but could achieve a strategic purpose. Its purpose may be to inflict mass casualties or destroy high-visibility enemy systems in order to weaken the enemy's national will to continue the intervention.

3-109. Likewise, victims of terror tactics may not be able to tell whether they were attacked by actual terrorists (independent or State-sponsored) or by insurgents, drug or criminal organizations, or SPF using terror tactics—but the results are the same. From the State's point of view, it can exploit the effects such attacks have on the enemy's tangible capabilities and/or his intangible centers of gravity, regardless of whether the State planned and carried out the attack or was merely able to capitalize on it and reap the benefits of someone else's action. Even when the State *is* responsible, there is opportunity for plausible deniability.

## TIMEFRAME

3-110. Strategic operations occur continuously, from prior to the outbreak of war to the post-war period. They can precede war, with the aim of deterring other regional actors from actions hostile to the State's interests or compelling such actors to yield to the State's will. Once war begins, they are generally conducted concurrently with the other components of the strategic campaign (regional, transition, and adaptive operations). What the various instruments of power do and which ones dominate in strategic operations at a given time depends on the same circumstances that dictate shifts from regional through transition to adaptive operations. Therefore, Chapters 4 through 6 will further discuss strategic operations in conjunction with the three other strategic-level courses of action.

3-111. The State is always applying its diplomatic-political, informational, and economic instruments of power. Even in peacetime, the military plays an important role. The very presence of the State's military power, which overmatches that of its regional neighbors, gives the State leverage and influence in regional affairs. Peacetime military engagement (PME) is another tool for expanding the State's influence. PME encompasses all peacetime programs and training exercises that the State's Armed Forces conduct to shape the international environment, open communications and improve mutual understanding with other countries, and improve interoperability with allies and potential allies. The State can also foster military or economic cooperation based on historical relationships. Thus, it may be possible for the State to achieve its strategic goals without ever resorting to armed conflict.

3-112. In wartime, strategic operations become an important, powerful component of the State's strategy for total war. They occur concurrently with regional, transition, and adaptive operations and can change the course of other strategic-level courses of action or even bring the war to an end.

3-113. Strategic operations may continue even after termination of the armed conflict. If the State succeeds in defeating the extraregional force or at least forces it to withdraw from the region, this victory enhances the State's status both regionally and globally. It will take advantage of this status to pursue its strategic goals. Should the State lose this war as judged from conventional political or military standards, but still survive as a nation and regime, it may be able to claim victory.

## STRATEGIC INFORMATION WARFARE

3-114. An important component the State's strategy for total war is the conduct of *information warfare* (IW), which the State defines as specifically planned and integrated actions taken to achieve an information advantage at critical points and times. The goal is to influence an enemy's decision making through his collected and available information, information systems, and information-based processes, while retaining the ability to employ friendly information, information-based processes, and systems.

3-115. Despite the fact that the State refers to it as "warfare," IW exists in peacetime as well as during war, and it is not just a military function and concept. In the context of total war, IW encompasses all instruments of national power. In peacetime, IW involves struggle and competition, rather than actual "warfare," as states and non-state actors maneuver and posture to protect their own interests, gain an advantage, or influence others. During times of crisis and war, IW activities continue and intensify. Defensive IW measures are more strictly enforced, while some of the more offensive elements of IW come to the fore. Even the subtler elements may become more aggressive and assertive.

3-116. Thus, the State applies IW across all levels of government activity, at every level of conflict, and in peacetime interactions with other actors. Because of its significance to the overall achievement of the State's national security strategy, IW at the strategic level receives special attention.

### THE STRATEGIC DIMENSION

3-117. *Strategic information warfare* (SIW) is the synergistic effort of the State to control or manipulate information events in the strategic environment, be they political, economic, military, or diplomatic in nature. Specifically, the State defines SIW as any attack (digital, physical, or cognitive) against the information base of an adversarial nation's critical infrastructures.

3-118. The ultimate goal of SIW is strategic disruption and damage to the overall strength of an opponent. This disruption also focuses on the shaping of foreign decision makers' actions to support the State's strategic objectives and goals. Perception management activities are critical to SIW. The State attempts to use all forms of persuasion and global media to win the "battle of the story."

3-119. SIW can undermine an extraregional power's traditional geographic sanctuary from strategic attack. SIW is not confined to a simple zone of territory, but can extend globally to encompass attacks on an opponent's homeland or the homelands of various military coalition members.

3-120. In addition to using all its own assets, the State will seek third-party actors or outside resources to support its overall information strategy. The range of actors can include digital mercenaries, individuals sympathetic to the State, terrorist organizations or individual terrorists, and criminal organizations. The State facilitates these shadow networks as necessary and continuously cultivates and maintains them during peacetime.

## ELEMENTS OF IW

3-121. Integrated within IW doctrine are the following elements:

- **Electronic warfare (EW).** Measures conducted to control or deny the enemy's use of the electromagnetic spectrum, while ensuring its use by the State.
- **Computer warfare.** Measures ranging from unauthorized access (hacking) of information systems for intelligence collection purposes to the insertion of destructive viruses and deceptive information into enemy computer systems. Such attacks focus on the denial, disruption, or manipulation of the infrastructure's integrity. SIW typically targets critical nodes or hubs, rather than targeting the entire network or infrastructure.
- **Deception.** Measures designed to mislead the enemy by manipulation, distortion, or falsification of information to induce him to act in a manner prejudicial to his interests.
- **Physical destruction.** Measures to destroy critical components of the enemy's information infrastructure.
- **Protection and security measures.** Measures to protect the State's information infrastructure and to deny protected information to other actors.
- **Perception management.** Measures aimed at creating a perception of truth that best suits State objectives, using a combination of true, false, and misleading information targeted at the State's own citizens and/or external actors. This element is crucial to successful SIW. The State is continuously looking for ways to sway international opinion in its favor or impact critical foreign strategic decision makers.
- **Information attack (IA).** Measures focused on the intentional disruption of digital information in a manner that supports a comprehensive SIW campaign. Information attacks focus exclusively on the manipulation or degradation of the information moving throughout the information environment.

3-122. The seven elements of IW do not exist in isolation from one another and are not mutually exclusive. The overlapping of functions, means, and targets makes it necessary that they all be integrated into a single IW plan. At the national level, this is known as the strategic information warfare plan (SIWP). However, effective execution of SIW does not necessary involve the use of all elements in conjunction. One element may be all that is required to successfully execute as SIW action or a supporting action at the operational or tactical level. The use of each element or a combination of elements is determined by the overall situation and specific strategic goals.

## INFORMATION ENVIRONMENT AND INFRASTRUCTURE

3-123. Most of today's information environment is outside of military control, making it harder to regulate, dominate, or protect. While neither the State nor its opponents can control the global information environment or global information infrastructure (GII), they must prepare to operate within it. The GII is defined as the worldwide interconnection of communications networks, computers, databases, and consumer electronics that make vast amounts of information available to users. The GII encompasses a wide range of equipment, including cameras, scanners, keyboards, facsimile machines, computers, switches, compact disks, video and audio tape, cable, wire, satellites, fiber-optic transmission lines, networks of all types, televisions, monitors, printers, and much more. The personnel who make decisions and handle the transmitted information constitute a critical component of the GII. Within the GII, various countries have their own the national information infrastructures (NIIs) and defense information infrastructures (DIIs). The GII provides global communications among NIIs and DIIs.

3-124. The NII is the physical and virtual backbone of a nation. It is composed of multiple critical infrastructures which support the following:

- Transportation and energy systems.
- Defense installations.
- Banking and financial assets.
- Water supplies.
- Chemical plants.
- Food and agricultural resources.
- Police and fire departments.
- Hospitals and public health systems.
- Government offices.

Critical infrastructures are those information and communication assets, systems, and functions so vital to a nation that their disruption or destruction would have a debilitating effect on national security, economy, governance, public health and safety, and morale.

3-125. Thus, an adversary's NII is the primary target for SIW. All critical information infrastructures that support the integrity of an adversary's social, political, economic, and military domains can be potential targets for disruption or entry points for perception management activities. The potential weaknesses in a technology-based society are numerous and complex because of the size and interconnectivity of such infrastructures. The State would attempt to take full advantage of such vulnerabilities and use the complexity of such infrastructures to hide or disguise its own involvement in SIW targeted against them.

3-126. The DII is defined as the shared or interconnected system of computers, communications, data applications, security, people, training, and other support structures serving an actor's defense needs. The DII connects computers used for mission support, command and control, and intelligence through voice, telecommunications, imagery, video, and multimedia services.

3-127. The interaction of the GII, NIIs, and DIIs introduces multiple actors into the information environment, increases vulnerabilities and dependencies, and creates many legal issues. This interaction compresses and blurs the distinctions among tactics, operations, and strategy. For example, images of tactical military actions, disseminated by the media, are likely to influence strategic decision makers or the populace. The State would constantly attempt to "spin" any conflict or situation to its advantage. Thus, perception management is a critical piece of SIW. The planning and implementation of SIW emphasizes increased use of psychological warfare and deception designed to manipulate public opinion, coupled with attacks against an opponent's centers of gravity.

## NCA STAFF RESPONSIBILITY

3-128. The NCA is responsible for the determination and articulation of the State's strategic goals. The Strategic Integration Department (SID) then develops a strategic information warfare plan (SIWP) to support the national security strategy. The SIWP establishes the overarching information strategy for achieving the State's strategic goals. Once approved by the NCA, the SIWP becomes a vital part of the national strategic campaign plan.

3-129. The SID is responsible for integrating all instruments of power—including the informational—across ministry lines. Indeed, the informational element is the instrument of power that most affects all State ministries and therefore requires the greatest integration effort. Because of this, the SID has a special Strategic Information Warfare Planning Office (SIWPO) dedicated to reviewing and integrating information-related plans of all State ministries, both military and civilian. The SIWPO can directly task information- or IW-related elements of any ministry to support the SIWP.

3-130. In the planning and execution of the SIWP, the SIWPO establishes priorities and assigns responsibilities for all State ministries and coordinates the information- or IW-related plans of all State ministries. It particularly coordinates the plans and actions of the Ministry of Foreign Affairs, Ministry of Public Information, Ministry of Finance and Economic Affairs, Ministry of the Interior, and MOD. Each of these key ministries has primary responsibility for one of the four instruments of power and develops its own campaign plan as part of the unified national SCP. Thus, there is a diplomatic SCP, a public information SCP, an economic SCP, and a military SCP. However, because the informational instrument of power cuts across the other three instruments of power and all government ministries, overall responsibility for SIW does not reside just within the Ministry of Public Information. Rather, the SIWPO maintains control of the collective IW activities of the various ministries and integrates the information-related plans of all those ministries into a unified, comprehensive, national-level SIWP.

3-131. The SID maintains liaison offices in all important State ministries to coordinate civilian and military activities during peacetime, mobilization, and war. Each SID liaison office has a section focusing on activities related to information and IW within the respective ministry. This liaison ensures that information and IW activities within all ministries are in concert with the SIWP and are integrated with related activities of other ministries in order to produce a cohesive, synergistic effect.

3-132. In times of war, the SIWPO continues to coordinate with all government ministries for further development and modification of the SIWP. However, it works most closely with the MOD, specifically the General Staff, to ensure the development of the SIWP in concert with the military IW plan.

## MINISTRY RESPONSIBILITIES

3-133. All State agencies take measures to ensure effective control of information and the appropriate dissemination of information within the State government, to the State's populace, and to other actors in the international environment. Each State ministry is responsible for the political indoctrination of its personnel and the dissemination of government policy information among its subordinate elements. Information security is also a function within each ministry. Each of the four key ministries associated with the four instruments of national power has its own campaign plan, which includes an IW plan.

### Ministry of Defense (General Staff)

3-134. The MOD has a military IW plan that integrates the IW plans and activities of all service components and major force groupings. Within the military structure under the MOD, the General Staff acts as the executive agent for the NCA. There is a Chief of IW under the Intelligence Officer in the Intelligence Directorate of the General Staff—just as there is a chief of IW under the intelligence officer in all military staffs down to brigade level. Within those operational- and tactical-level staffs, the intelligence officer and chief of IW are responsible for ensuring that all IW actions undertaken at their levels are in concert with the overall military IW plan and the SIWP. As necessary, the Chief of IW in the General Staff can directly task each operational- or tactical-level chief of IW to support the SIW campaign.

3-135. In the General Staff, the Chief of IW handles IW functions that transcend service component boundaries. He reviews and approves the IW plans of all operational-level commands as well as any separate theater headquarters that might be established. He drafts the overall military IW plan that, upon approval by the Intelligence Officer, is forwarded to the Operations Directorate of the General Staff for inclusion in the military SCP. Once approved by the CGS, the military IW plan and the rest of the military SCP are forwarded to the SID for incorporation into national-level SIWP and the national SCP, respectively. During peacetime and preparation for war, the Chief of IW continues to review and refine the military IW plan, and the SIWPO does the same with the SIWP.

3-136. The Operations Directorate of the General Staff assigns liaison officers to all important government ministries. The Chief of IW may also assign liaison officers to such ministries in order to keep abreast of information- and IW-related activities there. These officers often colocate with the SID liaison offices in those same ministries.

### Ministry of Foreign Affairs

3-137. The Ministry of Foreign Affairs has its own internationally oriented IW plan that includes diplomatic information campaigns and public relations campaigns directed at other countries (both within the region and external to it). The purpose of such campaigns is to explain, promote, and gain external support for the State's foreign policy objectives. At the government-to-government level, the ministry's Diplomatic Activities Directorate conducts diplomatic information campaigns to promote and protect the State's goals and national interests and to extend the political influence of the State. The State consciously uses diplomatic channels of communication to inform foreign nations about its policies and create a favorable response. The ministry's Public Relations Directorate also conducts information campaigns (in this case called public relations campaigns) within the region and external to the region. The directorate produces public relations films and articles and coordinates events that extol the virtues of the State.

### Ministry of Public Information

3-138. The Ministry of Public Information has a public information SCP primarily targeting the State's own population or regional actors. This ministry is responsible for the control and appropriate dissemination of all political, diplomatic, economic, and military information to the public and the international audience.

3-139. The State Information Directorate is responsible for the control and dissemination of information within the State. This directorate coordinates with the SIWPO to encourage constructive public support for the State's policy objectives and to counter hostile attempts to distort and frustrate the State's policies and programs. The wide range of public information campaigns involves all of the internal media sources, which this directorate monitors and controls. Its reviews and censors all information prior to authorizing its public release.

3-140. The Political and Cultural Information Directorate directs an information campaign aimed at producing government-approved information for dissemination to the all the State's residents. It disseminates this information through schools, social organizations, and private and government workplaces.

3-141. The International Information Directorate plays a major role in perception management activities to influence international political and public opinion regarding the State and its actions. This directorate maintains tight control of information released to the international world. It attempts to influence the international media and the consumers of this information to ensure that they will actively support the policies and goals of the State.

### Ministry of Finance and Economic Affairs

3-142. The Ministry of Finance and Economic Affairs has an economic information campaign plan for the appropriate protection and effective dissemination of information on the economic situation and economic policy, both internally and externally. The strength and vitality of the State's economy provide the capacity to influence the economic behavior of other nations and can also influence those nations' foreign policy and military behavior in ways that would benefit the State.

### Ministry of the Interior

3-143. The Ministry of the Interior also has IW-related functions among its internal security forces, some or all of which are resubordinated to the SHC (MOD and General Staff) in wartime. The primary imperative for the internal security forces is preservation of the regime and all four instruments of national power. This includes protection and control of information.

3-144. As a nondemocratic nation, the State suppresses and manipulates its people's ideas and values to rally support for State policies and deter dissidence. When internal security forces crack down on dissidents, government-controlled media either do not report the incidents or manage public perception to put the blame on the anti-government group. Monitoring of subversive groups includes eavesdropping on communications of such groups and their known leaders. Internal security forces also monitor foreign organizations operating within the State to detect negative influences or find ways for State agencies to manipulate these organizations to support the State's national security objectives.

# Chapter 4

# Regional Operations

*Regional operations* are a strategic-level course of action against opponents the State overmatches in conventional power, including regional adversaries and internal threats. They include conventional, force-on-force military operations, along with the application of the State's other instruments of national power.

Chapter 3 discusses regional operations in general terms within the context of the State's overall framework for implementing its national security strategy. This chapter outlines in more detail the goals of regional operations and how the State employs its diplomatic-political, informational, economic, and military instruments of national power in pursuit of those goals. It also describes how the State's strategic operations complement regional operations.

## STRATEGIC GOALS

4-1. The national security strategy is designed to achieve one or more specific strategic goals within the State's region. Therefore, it typically starts with actions directed at an opponent within the region—an opponent that the State overmatches in conventional military power, as well as other instruments of power.

4-2. The State's primary strategic goal in conducting regional operations is to expand its sphere of influence within the region. The specific goal might be territorial expansion, economic expansion, acquisition of natural resources, or the protection of a related minority population in a neighboring country. When the State no longer believes that the status quo offers a means to achieve its goal, it conducts regional operations.

4-3. To seize territory and otherwise expand its influence in the region, the State must destroy a regional enemy's will and capability to continue the fight. It will attempt to achieve strategic decision or achieve specific regional goals rapidly, in order to preclude outside intervention. In conjunction with military means, it can use political-diplomatic, informational, and economic instruments of power against regional threats and to prevent or preclude extraregional intervention.

4-4. During regional operations, the State must still contend with internal opponents, as well as the looming possibility of extraregional intervention. It will try to predict what extraregional actors may try to do to thwart its efforts, when they will do so, and how. It will try to prevent deployment of extraregional forces or an extraregional actor's intervention by other than military means. The State will also start planning for potential transition operations and even adaptive operations, in case these become necessary.

## DIPLOMATIC-POLITICAL MEANS

4-5. The State uses political means to maintain internal stability, while using diplomatic means to counter regional and extraregional threats. During regional operations, the State must control not only its own population, but also that of any conquered territory. It must continue to identify and neutralize internal dissident groups, paying particular attention to groups within the State that have ethnic, religious, and cultural backgrounds similar to those of the people of the country the State is fighting. The State closely monitors these groups to ensure that the groups' dissidents do not negatively affect the State's operations. It must also keep dissident groups from exploiting the state of total war or specific events for their own purposes.

4-6. In general, the State increases surveillance of its own citizens, monitoring their attitudes and activities. For example, soldiers who were conscripted into a peacetime army may now find themselves fighting a war they do not support. Particularly if they are fighting a people demographically related to them, such soldiers' questionable loyalty may threaten the State and its military operations. Families of soldiers killed in a war these families do not support may stage demonstrations against the State. Riots, sabotage, internal terrorism, work stoppages, and factory shutdowns can also threaten the State. Should regional operations negatively affect the supply of food and manufactured goods within the State, citizens who would normally remain loyal to the State may turn against it.

4-7. When necessary, the State uses internal security forces to control the local populace.[1] By monitoring known resident dissident leaders, the State hopes to determine what role, if any, they play in anti-State demonstrations and activities. The State will round up well-known dissident leaders and jail them to prevent them from agitating the masses and recruiting additional support. The State can establish curfews and use roadblocks, checkpoints, and sentries at borders to prevent other dissidents from entering the State from neighboring countries. The State can infiltrate groups that advocate the overthrow of the State government or do not support the State's goals. The State can extend the duration of conscripted military service indefinitely and not allow men to leave military service until a date to be determined by the State. The State actively seeks and rounds up those who disobey conscription notices. When necessary, the State can also use regular units of its Armed Forces against disloyal elements of its own population.

4-8. The State's internal security forces are responsible for identifying and neutralizing subversive elements regarded as threats to the regime. Both violent and peaceful groups are included in this category. Plainclothes agents investigate and monitor such groups and infiltrate their ranks. The internal security forces are engaged in intelligence-gathering and counterintelligence activities, employing an extensive human intelligence (HUMINT) network. Its agents are recruited from the most politically reliable segments of the State population. These agents not only gather information, but also focus

[1] When subordinate to the Ministry of the Interior in peacetime, the internal security forces have a primarily political role, although many of these forces are organized and equipped along military lines. Among dissident groups, these forces are sometimes called the "secret police." In wartime, they take on more of a military role, but at least some of them retain their politically-oriented mission.

the subversive element in a direction that the regime can control. The internal security forces also eavesdrop on all important telephone lines and installations dangerous to the regime. Through tapping mobile-phone connections, and especially through private radio stations, the internal security forces monitor the activities of dissident and subversive groups.

4-9. During regional operations, the internal security forces continue to operate their HUMINT network within the State and expand this network into enemy territory as the cornerstone of its reconnaissance, intelligence, surveillance, and target acquisition (RISTA) capability. This HUMINT network monitors and collects information on foreign organizations operating within the State's sphere of influence. These include not only foreign spies, but also foreign-based corporations doing business in the region, as well as nongovernmental and private volunteer organizations (NGOs and PVOs) that offer humanitarian assistance. Government agents infiltrate NGOs and PVOs to determine these organizations' agendas and how it can manipulate these organizations to support the State's goals and objectives.

4-10. During regional operations, the State uses diplomatic means with the primary purpose of preventing other actors from entering conflicts against the State. Diplomats will be active outside the country, forming alliances and seeking promises of neutrality or active support from neighboring states. This support may be political, military, or economic, and may be either overt or covert.

4-11. The State will seek permission to use forward staging areas, airfields, ports, and lines of communication (LOCs) located within regional countries, as well as to use regional airspace for hostile action or overflight. At the same time, it will deny ground transit and landing rights within the State to those countries that do not support its activities. It will also deny overflight within its borders by nations hostile to State goals. The State will ask neighboring countries to monitor their own internal dissident groups that may assist the State's dissidents in negatively affecting State operations.

4-12. At the conclusion of successful regional operations, the State would try to consolidate its regional power and stabilize the regional balance of power in its favor. It would also strive to gain global recognition of its expanded power status.

## INFORMATIONAL MEANS

4-13. The State conducts an internal information campaign aimed at maintaining and strengthening the national will. This campaign tries to give the State's people the impression that the government is keeping them informed, give them a positive attitude about the national leadership, and paint a picture of a common foe they can universally hate. The overall goal is to give the entire country a common focus.

4-14. The State's Ministry of Public Information continues to maintain tight control of national and local communications and internal media. It can manipulate the media to undermine support of internal adversaries and to strengthen the general population's support of State policy in internal and external matters. It ensures that newspaper editorials always support State actions. It can stage rallies by the local populace for the purpose of showing public support for State's actions. It would mandate heavy media coverage of such rallies and other positive events. It can crush negative rallies organized by the local populace and not allow the media to cover these rallies. The State attempts to focus subversive elements in a direction that the State can control.

4-15. The State conducts political education sessions at local factories and throughout the industrial base. It also continues indoctrination of youths within the Youth Corps. Children too young to join the Youth Corps receive indoctrination at school.

4-16. State-controlled media can project an image of the President as a national hero. His heroic status may be the result of past military victories or success in the ongoing strategic campaign. To those who share a common ethnic, cultural, or religious background with the President, the media may portray him as a champion of those causes. To all, he will be portrayed as a champion of State nationalism and expansionary goals.

4-17. In an attempt to show that volunteering for military duty is honorable and widespread, the State-controlled media can highlight stories about youth who voluntarily enlist. The State rewards families whose sons volunteer for military service. Presentations of these rewards to families are nationally publicized. Military retirees and others who had finished mandatory military service and returned to the civilian workforce, receive special recognition from the State when they volunteer for additional military service.

4-18. During regional operations, the Ministry of Public Information also continues to maintain tight control over access to international communications and international media. The State can confiscate radios owned by amateur radio operators communicating with and sharing news information with outside sources. Internet dial-up capability will be restricted to State officials only. Libraries and public places that allow the public to use the Internet will no longer be allowed to do so. The intent is to ensure that the populace is receiving positive messages about the State and its regional operations.

4-19. The State intensifies its efforts to control and manipulate international media. It will attempt to get foreign media to print favorable articles and run favorable television news spots about State activities. It can block access to international media's satellite transmissions critical of State operations; it controls the infrastructure (satellite downlink and hub) within the State and occupied territory and thus can deny distribution to the public by simply not broadcasting media received from the satellite. It can censor all newspapers imported into the State. Only newscasts and stories expressing favorable views of State operations are allowed to circulate.

4-20. State-controlled media will interview citizens of invaded territories and coerce them into discussing how much better off they are since State occupation, and the State will use these interviews as part of an information campaign. The State will take over radio and television stations, as well as print media, in the invaded territory and will use these media sources to present State political views to invaded peoples. It will identify, locate, and destroy underground newspaper sources in the invaded territory.

## ECONOMIC MEANS

4-21. The economic impact of regional conflict on the State depends on the level of resistance from the State's regional opponents and reaction from other countries. Even if a regional adversary has the military capability to conduct retaliatory strikes onto State territory, including economic targets, the State can avoid or degrade that capability by having civilians at such sites or by making it known that strikes would also destroy production of civilian goods. If the regional opponent causes some damage or outside actors impose economic sanctions, the impact on the State's economy may be greater.

4-22. In the best case, the State's economy and industrial base may suffer little or no damage from the war. The State may actually increase production in a rush to increase national wartime stockpiles of key military and civilian supplies and material prior to any extraregional intervention that could destroy production facilities or disrupt imports. Whether the State suffers damage to its domestic production capability or not, it may increase production at factories that the State or its ruling elite own in foreign countries. This can facilitate further stockpiling before extraregional intervention can close down those factories or the LOCs to them.

4-23. The State may use its economic leverage within the region in order to force neighboring countries to become its allies or at least to remain neutral in the conflict. It actively seeks economic support from regional actors and also searches for new sources of arms and other goods required to conduct sustained operations. It can boycott all goods and services exported from the country whose territory it has invaded, and it will attempt to convince other regional actors to do the same. Likewise, it will boycott economic goods and services from regional and extraregional actors not supporting State goals.

4-24. Once the State has totally subjugated the invaded territory, it will take over local factories and all means of industrial output within the territorial boundaries. It may also confiscate foreign-owned industries and assets within the State or the occupied territory.

4-25. In a worst-case situation, the State's internal resources might be strained, particularly if the State becomes subject to regional or international boycotts on its own imports and exports. The production of war goods might be at the expense of goods normally produced for civilian consumption. If necessary, the State is prepared to ration goods, including food and gasoline. It can appropriate large vehicles necessary for transport of military supplies and equipment. It can also appropriate privately-owned sea and inland-water transport capabilities. In order to increase industrial production to support the total war effort, the State may establish mandatory production quotas and extend workdays in factories. If necessary, factories may bring back retired workers to assist with factory operations.

## MILITARY MEANS

4-26. The State's choice to use military power indicates that the use of other means alone proved insufficient to reach the strategic goal, or that the State has elected to seize an opportunity that it can exploit militarily. Even in the context of regional operations, a strategic campaign may include several combined arms, joint, and/or interagency operations. If the State succeeds in forming a regional alliance or coalition, operations may also be multinational. Due to its military superiority over the regional adversary, the State is able to pursue primarily offensive military operations. It is also prepared to use military means against internal and possible extraregional threats.

4-27. The State's investment strategy for its Armed Forces focuses on regional dominance, so that regional objectives are achievable with existing military capabilities. Thus, it maintains large forces, in comparison to its regional neighbors, and pursues a program of selective modernization, insertion of new technology into older systems, and investment in a few high-cost, high-payoff systems that provide it a technological niche.

4-28. The State's military forces are sufficient to overmatch any single regional neighbor, but not necessarily an alliance or coalition of neighboring countries. They are certainly no match for the forces an extraregional power can bring to bear. Thus, the State seeks to exploit its numerical and technological overmatch against one regional opponent rapidly, before other regional neighbors or an extraregional power can enter the fight.

4-29. Regional operations are multiservice operations that include Army, Navy, Air Force, Strategic Forces, Special-Purpose Forces (SPF), and Internal Security Forces. Therefore, the State fields a large Army ground combat force—supported by the Air Force, Navy, and SPF—to seize territory or defeat a regional opponent. Strategic Forces may support these operations by delivering conventional weapons or nuclear, biological, or chemical (NBC) weapons, depending on the degree of escalation. In addition to their peacetime missions, the paramilitary Internal Security Forces can help control the population in territory the State seizes or engage enemy forces that invade State territory.

4-30. The State's military goal during regional operations is to destroy its regional opponents' military power in order to achieve specific ends. The State plans regional operations well in advance and executes them as rapidly as is feasible in order to preclude intervention by outside forces. Still, at the very outset of these operations, it lays plans and positions forces to conduct access-control operations in the event of outside intervention. Extraregional forces may also be vulnerable to conventional operations during the time they require to build combat power and create support at home for their intervention.

4-31. Plans and preparations for regional operations typically include efforts to ensure—

- The successful penetration of enemy defenses.
- Dependable fire support.
- Simultaneous deep operations throughout the region or theater that conform to the strategic campaign plan.

- Rapid exploitation.
- Countermeasures against enemy strike and reconnaissance systems.
- Preparation for access-control operations.

## COMMAND AND CONTROL

4-32. The State's overall command and control ($C^2$) concept is that of centralized planning and decentralized execution. Planning occurs from the top down. Accordingly, the higher commander's decision for the base course of action includes four essential elements: objective, opportunity, method, and end state.

4-33. Clearly stating *objectives* from the strategic level to the tactical level assures that all units understand the purpose for missions and the desired outcome. Objectives are linked at each level to achieve the State's purposes.

4-34. *Opportunity* stems from a common understanding of the conditions. In the State Armed Forces, plans identify branches and sequels that may take advantage of opportunity at all levels. Strategic-level actions create opportunity for operational action, and operational-level actions create opportunity for tactical action. The essence of the element of opportunity is that the State Armed Forces encourage initiative by subordinate commanders consistent with the objectives and end states identified during planning. Confronted by the possibility of intervention by more capable forces, the State relies on its commanders to act in the absence of orders, if required, to mitigate State limitations and vulnerabilities in its $C^2$ architecture.

4-35. Specifying *method* assures an effective allocation of resources and supports coordination in the absence of orders from higher headquarters. Finally, clearly articulating the desired *end state* assures that all State Armed Forces elements will execute toward a common goal, whether or not their communications systems are fully functional.

4-36. Detailed planning starts from the top down. The higher commander states the mission of subordinates in broad terms, accompanied by his concept of operations, which contains the essential elements of his plan. Thus, in the event circumstances change, a subordinate who is familiar with his superior's concept can adapt his efforts to ensure his unit contributes to the overall goal.

4-37. Initiative and creative approaches are the main criteria. In the State view, initiative consists of intelligent anticipation, or at least correct interpretation, of the higher commander's intent and the effective implementation of it without detailed guidance. It involves the flexible organization and employment of forces to react speedily, without waiting for direction, to meet unexpected changes in the operational and/or tactical situation. This enables the State to accelerate the pace of its operations and decision making so that it is able to function effectively at a higher tempo than its regional opponents can undertake.

4-38. Centralization of control gives the Armed Forces flexibility in the employment of resources to meet the overall goal of the strategic campaign. It ensures unity of views on the management of forces. Above all, it is essential to the control of weapons of mass destruction (WMD) and precision weapons. It is also important in the management of long-range fires and air defense operations.

## INTERNAL SECURITY FORCES

4-39. In wartime, the internal security forces of the Ministry of the Interior are resubordinated to the Supreme High Command (SHC) and act as a sixth service component of the Armed Forces. At that time, the formal name Internal Security Forces applies to all forces resubordinated from the Ministry of the Interior to the SHC, and the General Staff controls and supervises their activities. Especially during wartime, dissidents must be neutralized by all means necessary in order for the State to maintain firm control of its population. When the State occupies enemy territory during war with a regional neighbor, it can deploy some of these forces into the occupied territory to establish similar control over the local population. In addition to the paramilitary Internal Security Forces, the State may employ other elements of the Armed Forces to deal with internal threats, when necessary.

4-40. In turn, the State may use any or all of its Internal Security Forces to deal with enemy military forces. They can engage enemy special operations forces operating in denied areas or even enemy conventional forces, when operating in cooperation with another organization from the State's Armed Forces. Many elements of the Internal Security Forces are organized along military lines and equipped for combat. For instance, the National Security Forces have light weapons, some heavy weapons, and armored vehicles. Thus, the State can use these forces against regional or extraregional forces that invade State territory. Because some special police units are equipped with heavy weapons and armored vehicles, they have the combat power to conduct defensive operations if required. The State Security Directorate has its own highly-trained SPF teams, equipped to conduct direct-action missions in the enemy's rear. Other paramilitary organizations with somewhat lesser capabilities include constabulary forces, territorial militias, and home guards. Within the State, there are also private security organizations belonging to business enterprises, industry, or local ad hoc groups.

## ARMY

4-41. The Army is large and modern, in comparison to the ground forces of the State's regional neighbors. Therefore, it is capable of offensive operations against such opponents, relying heavily on its armor, mechanized infantry, and airborne forces. However, it also has large infantry forces suitable for operating in the urban environments and other complex terrain that dominate portions of the region. Its fire support is adequate to dominate regional adversaries. Army SPF can support the ground forces at the operational level and conduct reconnaissance and direct action to the opponent's operational depth.

4-42. Army ground forces have two purposes: to destroy other military forces or to seize terrain. For these purposes, Army forces may attack along a suboptimal approach to exploit an enemy vulnerability or to achieve surprise. To maintain a high tempo of operations and reach key targets, Army forces often accept the risk of bypassing pockets of resistance.

4-43. Offensive operations during the course of regional operations attempt to achieve strategic political or military decision by destroying the enemy's will and capability to fight. The Army may try to bring this about by destroying the $C^2$ and logistics systems the enemy needs for continued operations.

## NAVY

4-44. The Navy is more than adequate to operate against navies of regional opponents and overwhelm them. Its primary means of dominating regional navies are small fast-attack craft armed with antiship missiles. These craft, along with ground-based antiship missile units, submarines, and mines, also give the Navy the ability to challenge or control access to the region by sea.

4-45. The Navy enables the State to project power within the region. It can project ground forces or naval infantry up to brigade size within the region. If necessary, the State can appropriate privately-owned ships to supplement its transport capability. Naval infantry forces can operate independently for up to 30 days pending linkup with ground forces. Naval SPF can carry out reconnaissance in support of landings or conduct raids against critical targets.

## AIR FORCE

4-46. The State takes great pride in its Air Force, which is numerically and technologically superior to other air forces in its region. It invests in a few high-payoff systems and larger numbers of other aircraft that are less than state-of-the-art but still sufficient to dominate the regional airspace. It has transport aircraft capable of projecting airborne troops regionally. It can also insert its own SPF and those of other service components.

4-47. The Air Force also includes Air Defense Forces with which the State can successfully defend its airspace against regional opponents. Against such adversaries, the State may be able to use an integrated air defense system (IADS) that is centrally directed at the national level. The State's air defenses also provide the capability to challenge or deny air access into the region by outside forces, at least initially.

## STRATEGIC FORCES

4-48. The long-range missiles and rockets of State's Strategic Forces are primarily political tools for exerting influence in the region. When necessary, the State can use these systems with conventional warheads to strike key targets and affect the national will of a regional opponent. The threat of using these systems to deliver WMD is also an intimidating factor. Should any regional opponent use its own WMD capability against the State, the State is prepared to retaliate in kind. It is also possible that the State could use WMD against a regional neighbor as a warning to any potential extraregional enemy that it is willing to use such weapons.

4-49. Strategic Forces can use long-range missiles and even WMD to deny a regional opponent the use of urban and other complex terrain. This creates opportunities for operational forces to engage the enemy with fires and maneuver.

## SPECIAL-PURPOSE FORCES COMMAND

4-50. The State typically inserts SPF teams in advance of regional operations to conduct reconnaissance and direct action. The SPF can also support terrorism and insurgent operations in the region.

## AFFILIATED FORCES

4-51. The State may also have overt or covert affiliations with other forces that act in concert with the Armed Forces but are not actually part of them. These affiliated forces may be mercenaries, insurgents, terrorists, and drug or criminal organizations. Thus, regional operations could include State-sponsored terrorism or insurgency against a regional neighbor. Other affiliated forces may include cyber terrorists or hackers.

# STRATEGIC OPERATIONS

4-52. During the course of regional operations, the State uses strategic operations primarily in defensive ways, in order to prevent other parties from becoming involved in what it regards as purely regional affairs. At this point, use of any military means against parties not currently involved in the conflict would most likely have the opposite effect, causing them to become involved. Therefore, the State relies primarily on the diplomatic-political, informational, and economic means in a peacetime mode in relation to parties with whom it is not at war. For example, it may try to deny the rest of the world information on events in the region or to portray those events in a manner favorable to the State.

## Chapter 5

# Transition Operations

*Transition operations* are a strategic-level course of action that bridges the gap between regional and adaptive operations and contains some elements of both. The State continues to pursue its regional goals while dealing with the development of outside intervention that has the potential for overmatching the State.

Chapter 3 discusses transition operations in general terms within the context of the State's overall framework for implementing its national security strategy. This chapter outlines in more detail the goals of transition operations, the nature of the transition, and how the State employs its diplomatic-political, informational, economic, and military instruments of national power in pursuit of its strategic goals. It also describes how the State's strategic operations complement transition operations.

## STRATEGIC GOALS

5-1. The State conducts transition operations when other regional and/or extraregional forces threaten the State's ability to continue regional operations in a conventional design against the original regional enemy. Transition operations serve as a means for the State to adapt to the new situation and still pursue its overall strategic goal of regional expansion. At this point, another emerging strategic goal is to defeat outside intervention or perhaps still prevent it.

5-2. Transition operations serve as a bridge between regional and adaptive operations. The transition may go in either direction. The fact that the State begins transition operations does not necessarily mean that it must complete the transition from regional to adaptive operations (or vice versa). As conditions allow or dictate, the "transition" could end with the State conducting the same type of operations as before the shift to transition operations.

5-3. Usually, the State does not shift directly from regional to adaptive operations. The transition is incremental and does not occur at a single, easily identifiable point. Thus, a period of transition operations overlaps both regional and adaptive operations. The State plans and prepares for transition operations prior to being forced into adaptive operations; so, the transition can begin concurrently with regional operations. As the term "transition" implies, actions defy clear categorization and the progression is not easily discernable. Transition operations allow the State to shift gradually to adaptive operations or back to regional operations. In this fluid situation, transition operations serve as a pivotal point in a strategic campaign that could go either way. (See Figure 5-1.) At some point, the State seizes an opportunity to return to regional operations or it reaches a point where it must complete the shift to adaptive operations.

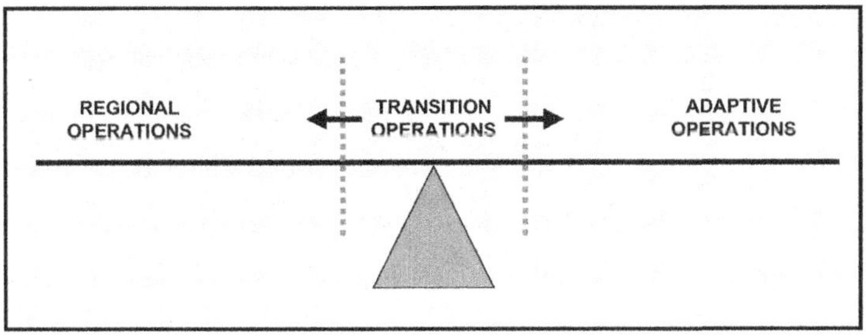


**Figure 5-1. Transition Operations**

5-4. Mostly defensive in nature, transition operations may include some offensive operations. If this combination of offensive and defensive actions is successful and the extraregional force is no longer a factor, the State may be able to transition back to regional operations without having to complete the shift to adaptive operations.

5-5. During transition operations, the State must decide whether to keep its forces in any territory it has occupied in a neighboring country or to withdraw them back to its home territory. The decision to stay or withdraw at this point may be based on the presence or absence of urban areas and/or other complex terrain suitable for sanctuaries and other adaptive measures in the occupied territory against an extraregional power with overmatch in technology and conventional forces. Sanctuary requires not only suitable terrain, but also a sympathetic or intimidated local populace. The State is also more likely to remain in the occupied territory if it has already achieved its strategic goal in regional operations or at least achieved major intermediate objectives leading toward that goal.

5-6. At the point of shifting to transition operations, the State still has the ability to exert all instruments of national power against an overmatched regional enemy and, indeed, may have defeated its original regional adversary. However, its successful actions in regional operations have prompted ether other regional actors or an extraregional actor to threaten to become involved in the conflict. The State will use all means necessary to preclude or put a quick end to such possible expansion of the conflict and to either consolidate its previous gains or conduct further operations against the original regional enemy using a more conventional design.

5-7. Transition operations can also buy time for the State's strategic operations to succeed. Meanwhile, strategic operations against the impending extraregional threat may resort to more offensive actions against intangibles and even carefully selected attacks on the enemy's tangible assets in order to target his strategic centers of gravity. For example, the State or its affiliated forces can attack staging areas and economic targets in the region or even in the enemy's homeland.

## DIPLOMATIC-POLITICAL MEANS

5-8. Because of the lack of a clear-cut division between regional and transition operations, political means of maintaining internal stability during transition operations are similar to the means used during regional operations. The types of internal threats remain the same, but are intensified by changing conditions. When conducting regional operations, the State was making progress toward a longstanding goal of regional hegemony and enjoyed a high level of public support. When extraregional intervention threatens to halt or reverse this progress, it also threatens internal stability. Faced with increasing unrest among local civilians, the State must be more concerned about internal dissidents and use its control mechanisms to crack down on them. However, the State's ability to deal with increased internal unrest is impacted by diversion of resources and instruments of national power to deal with external threats from regional and extraregional forces. When the transition is from adaptive back toward regional operations, there will also be a lack of clear distinction in the use of political means against internal threats.

5-9. As it does during regional operations, the State will continue to use diplomatic means to negotiate alliances and attempt to increase support internally and externally to the region. However, these negotiations take on more of a sense of urgency for the State. It may enlist the services and expertise of former or retired diplomats and politicians who have had experience working with various regional and extraregional actors and have established a rapport with the officials in these countries. During these negotiations, the State may use bribes or make promises that it has no intention of keeping. Against extraregional actors, it will initiate diplomatic actions to postpone, delay, or disrupt the mobilization and deployment of extraregional forces. Even after actual intervention begins, the State will seek diplomatic ways to prevent deployment of further forces.

## INFORMATIONAL MEANS

5-10. The State continues to exercise the informational instrument of power during transition operations in the same manner it uses during regional operations. In its internal information campaign, it will exaggerate enemy combat losses as compared to State combat losses. The State will not admit that it is overmatched by regional and/or extraregional adversaries, that it is in danger of losing the war, or that it is failing to make progress in achieving its strategic goal. The State will accentuate success and will put a positive spin on all news releases. The State's informational goal will be to convince its citizens that transition operations are necessary in order for the State to exploit the many gains it has already made.

5-11. In information campaigns targeting the international community, the State increases its emphasis on popularizing the State and its actions. Not wishing to appear as an aggressor, it attempts to convince the international community that it is conducting its primarily military campaign in order to help regional neighbors increase their standard of living and improve their way of life. If it is obvious that the State will be overmatched by the extraregional force that is about to intervene, the State may depict the intervening force as an unwanted aggressor involving itself in regional affairs in order to support its own selfish interests.

5-12. Against extraregional threats, the State begins to use more offensive forms of information warfare (IW). These include not only more aggressive information campaigns, but also information attack and perhaps physical attack, as long as there is opportunity for plausible deniability. As extraregional forces begin to deploy into the region, the State can use information attacks on enemy command and control ($C^2$) systems. The State continues to leverage international media to influence world perception and public opinion within the extraregional power's own populace. It can influence the extraregional force's operations by playing up in the media the fact that the enemy's standoff weapons sometimes strike innocent civilians.

## ECONOMIC MEANS

5-13. By the time the State enters transition operations, it should already have sufficient stockpiles of key military supplies and materiel to support sustained operations. Faced with economic sanctions and boycotts of imports, it now focuses on further stockpiling of critical civilian goods such as fuel, food, and clothing to satisfy the basic needs of the populace. Even with these stockpiles, the State may gradually begin rationing of civilian supplies and services as a hedge against future shortages. It assumes that the intervening extraregional force will be capable of striking its fixed production facilities. The State will seek to keep the enemy from using this capability by publicizing that there are large numbers of civilians at these sites or that such strikes would also destroy production of civilian goods.

5-14. During transition operations, if not before, the State begins to nationalize the industrial base. Since the wealthy families who control the State government also own most of the large factories and major business enterprises, the step of nationalizing these parts of the State's economy is a short one. The State may ask workers to work longer hours for lower wages, and use the money saved to further State goals. The State may increase production quotas and reward factories or workers who meet them while severely punishing those who do not. The State will increase interest rates on government-funded loans.

5-15. Prior to transition operations, the State monitored railway and port-of-entry operations closely. During transition operations, it may nationalize these assets. Thus it can dictate and control all railway schedules. By limiting passenger travel to official travel only, the State can use the railway lines to transport more goods to areas where they are needed to support the war effort. It can confiscate privately-owned container cars and use them for official business. The State will limit transfer of goods at all ports of entry to those goods necessary for the sustainment and conduct of the war effort.

5-16. The State will attempt to negotiate future payment for imported goods in order to decrease the cash flow out of the country. By calling in international debt and demanding debt payments from regional and extraregional actors, the State will also increase the cash flow into the State. In turn, the State may cease payments on national debt owed to other countries.

5-17. The State will use economic sanctions and pressure to prevent its regional neighbors from lending support or sanctuary for deployment of extraregional forces. It will also use the possibility of collateral damage to foreign-owned assets within the State as a means of deterring the extraregional power from using air campaigns or long-range weapons against the State.

## MILITARY MEANS

5-18. It is possible for transition operations to be hasty in nature—against an unexpected regional coalition or unexpected extraregional intervention. In this rather unlikely case, the short-term goal is preservation of military capability while seeking transition back to regional operations.

5-19. In most cases, the State will have anticipated extraterritorial intervention and will have thoroughly planned transition operations, as well as possible adaptive operations. The Armed Forces operate human intelligence (HUMINT) agents throughout the region and have some global intelligence-gathering capability against those nations the State perceives as the most likely potential enemies. When the transition is planned and deliberate, the long-term goal is preservation and application of military capability supporting the transition from regional to adaptive operations and eventually back to regional operations, if conditions support doing so.

5-20. During transition operations, military forces solidify gains made during regional operations. However, the central aim is to prevent or defeat outside intervention. Although military operations are primarily defensive in nature, limited attacks may continue. A combination of operational and tactical offensive and defensive actions help the State to control tempo. In some cases, the objective is to get inside the enemy decision cycle.

5-21. The State will use the time it takes the extraregional force to prepare and deploy into the region to change the nature of the conflict into something for which the intervening force is unprepared. The State tries to establish conditions that force the new enemy to fight at less than full strength and on terrain for which his forces are not optimized. It seeks to take advantage of urban areas or other complex terrain whenever possible, while controlling the enemy's access to such terrain. It plans operations to exploit the opportunities created by the presence of NGOs, PVOs, media, and other civilians on the battlefield.

5-22. Military forces in the immediate vicinity of the point of intervention move into sanctuary as opportunity allows, making use of existing $C^2$ and logistics. They conduct limited attacks to secure positions, protect flanks, and control access. They may attack vulnerable early-entry forces before the enemy can bring his technological overmatch to bear. Even at this stage, the State may be able to inflict politically unacceptable casualties that could cause the extraregional power to terminate its intervention.

5-23.During transition operations, State forces will plan and conduct sophisticated ambushes to destroy high-visibility enemy systems or cause mass casualties. These ambushes are not always linked to maneuver or ground objectives, but may have huge psychological and political impact by demonstrating the vulnerability. The State may use niche technology it has acquired to achieve technological surprise and limited-duration overmatch in specific areas.

5-24. The State can use long-range missiles to deny an extraregional opponent the use of urban areas or other complex terrain. This creates opportunities for operational forces to destroy key enemy systems with precision fires or to engage the enemy forces with fires and maneuver.

5-25. If access-control efforts are successful, in conjunction with transition operations, the State wants to stay in the regional operations mode. If attacks against enemy early-entry forces are successful, the State may go back to regional operations. If not, it moves toward adaptive operations. It wants to be able to go either direction.

5-26. Meanwhile, transition operations permit other key forces the time, space, and freedom of action necessary to move into sanctuary in preparation for a shift to adaptive operations. These forces preserve combat power and prepare to defend the State homeland, if necessary. Transition operations usually include mobilization of reserve and militia to assist in defending the State.

5-27. Transition operations may include several combined arms, joint, and/or interagency operations and, if the State is part of an alliance or coalition, may also be multinational. Even when operations are joint, the air and naval forces increasingly revert to defensive measures to preserve their capability. Ground forces or, more often, Special-Purpose Forces (SPF) conduct raids against logistics sites, lines of communication (LOCs), and vulnerable military targets in the region, along the routes to the region, and to the enemy's strategic depth. Occasionally, if the risks are worth the costs, the State attacks such targets by air and sea. It may also use long-range missiles or rockets to deliver conventional warheads or weapons of mass destruction (WMD) against these kinds of targets. Paramilitary forces, air defense forces, and precision attack can also play important roles. The State may also use long-range weapons or SPF to conduct attacks outside the theater, to divert enemy resources to protect politically or ecologically sensitive targets. Although these attacks are characteristically part of transition operations, they are also conducted during regional and adaptive operations if required. The purpose is to allow the enemy no sanctuary.

5-28. At some point, the State may conclude that it cannot deny entry or defeat the extraregional force by destroying his early-entry forces. The State then shifts its emphasis to completing the transition to adaptive operations as soon as possible, before the enemy can deploy overwhelming forces into the region.

## COMMAND AND CONTROL

5-29. During transition operations, the $C^2$ process becomes more complicated. Commanders must try to ensure that State forces prepare for three basically different functions on the battlefield. They must—

- Disperse forces and prepare sanctuaries for adaptive operations.
- Control operational forces as they either disengage from regional enemy forces or plan to engage extraregional forces in continuing transition operations.
- Plan for a return to regional operations.

The requirements for $C^2$ are different for delay actions or sanctuary-based operations than for offensive operations, which are much more mobile.

5-30. The principle of centralized planning and decentralized execution continues to apply in transition operations and becomes even more important. Planning at the highest levels, even during regional operations, has accounted for the possibility of intervention by an extraregional power. Now, decentralized execution allows subordinate commanders to react rapidly to this critical, but not unanticipated, change in the strategic situation without waiting for direction from higher headquarters. Thus, transition operations, although increasingly decentralized in their execution, proceed toward the overall goal of the strategic campaign.

5-31. As in regional operations, planning for transition operations includes objective, opportunity, method, and end state. In this case, the objectives are to preclude, delay, disrupt, or destroy outside intervention. State forces conducting the transition operations can either seize the opportunity to return to regional operations or provide other forces the opportunity to prepare for adaptive operations. Planned methods for transition operations are to—

- Provide adequate logistics support.
- Delay enemy deployment.
- Attack early-entry forces.
- Attack LOCs into the region and in the region.
- Develop air defense ambushes and pre-position forces to do so.
- Set the conditions to conduct regional or adaptive operations.
- Attack intervening forces to their strategic depth.

5-32. Transition operations do not lead to a single end state. The end state against a regional opponent may be consolidation and/or disengagement. If transition operations prevent intervention or destroy the intervening force, the result can set the conditions for a return to regional operations. If they merely delay or temporarily disrupt the intervention and the extraregional force is able to get sufficient force in the region to threaten eventual overmatch against the State, the desired end state is to permit other State forces to prepare for a shift to adaptive operations.

## INTERNAL SECURITY FORCES

5-33. The State becomes even less tolerant of internal dissent and unrest, since sanctuary requires cooperation of the local populace. Thus, it is more likely to use paramilitary or military force to neutralize such threats.

5-34. As during regional operations, the State may deploy Internal Security Forces into territory it has seized in a neighboring country. If the State cannot control the entire occupied area with regular military forces, it may use Internal Security Forces to control the local population.

5-35. The State may use any or all of its Internal Security Forces to engage enemy special operations forces operating in denied areas. The more heavily armed of these forces can engage enemy conventional forces, but usually in cooperation with another organization from the State's Armed Forces.

## ARMY

5-36. During transition operations, if not before, the State begins mobilization of militia for defense of the homeland. The militia forces belong exclusively to the Army and consist primarily of infantry. Thus, they are suited for securing LOCs and defending the State in sanctuary-based operations, but they are generally not capable of large offensive operations.

5-37. Regular Army ground forces or Army SPF conduct raids against enemy logistics sites, LOCs, and vulnerable military targets. Army forces may still be able to conduct operations in a conventional design against enemy early-entry forces. However, Army operations generally begin to move toward the less conventional applications associated with adaptive operations.

## NAVY

5-38. The Navy possesses the capability to challenge or control access to the region by the most modern navies at maritime choke points, using a combination of mines, shore- and sea-based antiship missiles, and submarines. The Navy can also insert naval infantry or Naval SPF to conduct raids against critical installations within the region. If necessary, the State can appropriate privately-owned ships or inland-water craft to supplement its capability to transport troops or supplies.

## AIR FORCE

5-39. At least during the early stages of intervention, the Air Force and its Air Defense Forces are sufficient to challenge extraregional air forces and prevent them from attaining air supremacy for a time. Thus, the State's air and air defense forces may prove effective in supporting access-control efforts during transition operations.

5-40. Within the region, Air Force transport aircraft may deliver airborne troops, Air Force SPF, and SPF belonging to other components. The State may also use privately-owned and commercial aircraft for this purpose. Air Force SPF can conduct raids against enemy air bases and other installations.

## STRATEGIC FORCES

5-41. The State can use the long-range missiles of its Strategic Forces to strike intermodal transportation nodes, and air and sea ports along the LOCs that any extraregional force might require for its deployment into the region. It may also use them to extend the conflict beyond the region, to affect the national will of potential opponents or members of enemy coalitions.

5-42. The Strategic Forces can strike enemy logistics sites, LOCs, and vulnerable military targets using missiles with conventional or WMD warheads. However, it is unlikely that the State would use those delivery means to employ NBC weapons inside or outside the region or theater prior to using the same units to deliver long-range conventional weapons. It also could use WMD tactically or operationally against troops in the field prior to using them strategically. However, once having crossed the threshold of using WMD, the State may use them against any high-payoff target within the range of its delivery systems. Positioning of the State's Strategic Forces units and supporting $C^2$ nodes could, if detected, serve as indicators of the State's intentions. Therefore, the State will seek to conceal the location of these forces and $C^2$ nodes.

### SPECIAL-PURPOSE FORCES COMMAND

5-43. During transition operations, the SPF Command can use its regional and global intelligence-gathering capabilities to the enemy's strategic depth. It can use SPF teams to conduct direct action attacks against ports, LOCs, and early-entry forces. The SPF can also support insurgent and terrorist operations to delay or disrupt the extraregional force's mobilization and deployment.

### AFFILIATED FORCES

5-44. During transition operations, the State can count on most of the same affiliated forces that aided it in regional operations. It may be able to form further affiliations with groups that were not originally sympathetic to the State's goals, but are willing to unite with the State against an extraregional power perceived as a common enemy. Affiliated forces can attack aerial and sea ports of debarkation (APODs and SPODs), staging areas, or economic targets in the region. Terrorists with global reach can even strike the homeland of the extraregional force.

## STRATEGIC OPERATIONS

5-45. The State will use strategic operations to limit or shape enemy coalitions. During transition operations, the military aspects of strategic operations become more aggressive. The State can try to slow down enemy mobilization by information attack. During deployment, it can disrupt the flow of enemy movement at his aerial and sea ports of embarkation (APOEs and SPOEs), as well as APODs and SPODs, by SPF direct action and information attack. It can physically attack enemy early-entry forces and their LOCs. It can also attack high-payoff targets such as high-technology communications nodes and other information systems.

5-46. Enemy force projection from a distant homeland affords the State the opportunity to mine nautical choke points and to establish antishipping ambushes. Air and sea ports, both in the enemy homeland and in the region, are vulnerable and convenient targets for terror tactics or sabotage in support of transition and adaptive operations. The State can also exploit local nationals employed as contractors and host-nation support for the extraregional force; the State can use them to collect intelligence or conduct sabotage.

## Chapter 6

# Adaptive Operations

*Adaptive operations* are a strategic-level course of action to preserve the State's power and apply it in adaptive ways against opponents that overmatch the State. Theoretically, this overmatch could be the result of an alliance among the State's regional neighbors, but it will more likely come from the intervention of an extraregional power.

Chapter 3 discusses adaptive operations in general terms within the context of the State's overall framework for implementing its national security strategy. This chapter outlines in more detail the goals of adaptive operations and how the State employs its diplomatic-political, informational, economic, and military instruments of national power in pursuit of those goals. It also describes how the State's strategic operations complement adaptive operations.

## STRATEGIC GOALS

6-1. Generally, the State conducts adaptive operations during the strategic campaign as a consequence of intervention from outside the region. If it cannot control the extraregional enemy's access into the region or defeat his forces before his combat potential in the region equals or exceeds its own, the State must resort to adaptive operations. The primary objectives are to preserve combat power, to degrade the enemy's will and capability to fight, and to gain time for aggressive strategic operations to succeed. However, the State will not cede the initiative. Even with the intervention of an advanced extraregional power, the State will employ military means so long as this does not either place the regime at risk or risk depriving it of sufficient force to remain a regional hegemon. Adaptive operations generally include the State's home territory, as well as the regional theater(s) in which the State has conducted regional operations.

6-2. Once an extraregional force intervenes with sufficient power to overmatch the State, the State's immediate goal is survival—as a regime and as a nation. However, its long-term goal is still the expansion of influence within its region. In the State's view, this goal is only temporarily thwarted by the extraregional intervention. Accordingly, planning for adaptive operations focuses on effects over time. The State believes that patience is its ally and an enemy of the extraregional force and its intervention in regional affairs.

6-3. It is the combination of the State's capabilities and strategy that make it believe it can take on the extraregional force and win. The State must make creative and adaptive use of existing technologies and forces. At the strategic, operational, and/or tactical levels, it employs conventional and unconventional forces in an adaptive manner, in close coordination with the strategic operations, to wrest the initiative from the opposition and achieve decisive operational and strategic results.

6-4. Internal threats remain much the same as in regional and transition operations, but are further intensified by the conditions that led to adaptive operations. As in transition operations, the State's ability to deal with increased internal unrest is impacted by diversion of resources and instruments of national power to deal with external threats from regional and extraregional forces.

6-5. The State believes that adaptive operations can lead to several possible outcomes. If the results do not completely resolve the conflict in the State's favor, they may at least allow the State to return to regional operations. Even a stalemate may be a victory for the State, as long as it preserves enough of its instruments of power to preserve the regime and lives to fight another day.

## DIPLOMATIC-POLITICAL MEANS

6-6. The State will conduct diplomatic and political activities similar to those during regional and transition operations. It becomes very difficult for the State to use political means to counter internal threats during adaptive operations. By the time the State has shifted to adaptive operations, it has suffered casualties and may not have reached its goal. The State's citizens may be rebellious and more difficult for the State to control during adaptive operations than during regional operations.

6-7. The State will expand and step up internal control measures implemented during regional operations. For example, it may make travel passes mandatory for travel inside and outside urban areas and around key strategic locations. It may establish curfews and close places of entertainment such as bars, theaters, and other gathering places where citizens could meet to form dissident organizations. It may even abolish all religious services and related activities.

6-8. By rewarding those citizens who continue to support the war effort, the State hopes to increase support from other citizens. It will fill jails with those who do not support it, and even provide monetary incentive for citizens turning in other citizens for unpatriotic activities.

6-9. The State continues to seek new diplomatic agreements and alliances that will allow it to expel the extraregional force that caused it to transition to adaptive operations. Unexpected alliances with the State can affect the extraregional force's dominance. The State also attempts to exploit the vulnerabilities of an enemy coalition.

## INFORMATIONAL MEANS

6-10. The State continues to censor and manipulate the media. It may change and modify its basic themes, but the final essence of the information campaign remains the same: to maintain or expand internal, regional, and international support for State operations.

6-11. During adaptive operations against an extraregional enemy, State-controlled media continue to project an image of the President as a national hero. The media seek to add to his heroic status. This image is important to impress on the general populace, so that it will continue to follow the President in hardship. Otherwise, the populace may not have the same patience as the State leadership or maintain the same willingness to sacrifice over time.

6-12. In its internal information campaign, the State will continue to exaggerate its gains and downplay the extent of its losses. It will continue to extol benefits of continuing the strategic campaign. It will tell citizens that its strategic campaign is a success and will continue to be an even greater success if the citizens dedicate more time and effort to the State's cause. It will continue to use the citizens of the invaded territory as pawns in its information campaign, which also will exploit enemy deserters, as well as prisoners of war from extraregional forces.

6-13. The State uses perception management and other tools to attack the enemy's will to fight or otherwise continue its intervention, and to manipulate international opinion. If it still occupies territory of a neighboring country, it also tries to turn the populace there against the intervening extraregional force.

6-14. During adaptive operations, the informational element of strategic operations continues to attack the intangible components of the enemy's efforts against the State. It targets those elements that can most affect factors such as enemy soldiers' and leaders' confidence, political and diplomatic decisions, public opinion, the interests of private institutions, national will, and the collective will and commitment of alliances and coalitions.

6-15. The extraregional force and its individual soldiers tend to lack cultural awareness of the region and will thus make mistakes that unintentionally offend the local populace. The State will publicize these unintentional offensive actions in order to solidify its own people's hatred of the outsiders or to turn regional neighbors against them.

6-16. The State can create or manufacture humanitarian crises to embarrass the enemy or force him to divert resources to correct the real or manufactured crises. It can also stage incidents, riots, and demonstrations for media attention and to delay or disrupt enemy movement.

## ECONOMIC MEANS

6-17. The State will continue tighten the internal economic controls it implemented previously. It will nationalize all small, private factories that had not previously come under its control. The State may close schools in order to allow children to join the workforce and assist in the production of war materials. The presence of children in the factories may deter the enemy from bombing them. The State will authorize release and use of equipment, energy sources, and goods that are stored in national wartime stockpiles. However, it will normally impose or increase rationing of civilian supplies and services.

6-18. The State continues and intensifies economic sanctions and pressure on other nations. It will look for new or expanded lines of communication (LOCs) that open up routes closed by intervention of extraregional forces. If diplomatic sanctions are not imposed, the State will continue to produce, transport, and stockpile key military and civilian goods from State-owned factories based in other countries. It will seek to transport supplies and materiel under foreign-flagged commercial transportation, as well as through the supply networks of affiliated forces.

## MILITARY MEANS

6-19. Adaptive operations occur as a result of an extraregional power intervening with sufficient forces to thwart the State's original offensive operations in the region. Adaptive operations are often more defensive in nature than were regional or transition operations. They may focus on preserving forces, retaining gains in the region, and defending the State's home territory. The State's forces disperse to the extent their command and control ($C^2$) allows and conduct decentralized operations in both offense and defense. The State views adaptive operations as temporary in nature, serving as a means for the State to return to regional operations.

6-20. Once an extraregional power commits forces in the region, State forces will not avoid battle. They will seek it often, but on their own terms. Battles will occur at a place and time of the State's choosing and will involve dispersed maneuver, precision fires, and simultaneous actions by all services of the Armed Forces as well as affiliated forces.

6-21. Adaptive operations are often sanctuary-based. Sanctuaries are areas that limit the ability of an opponent to apply his full range of capabilities. The State's forces can use physical and/or moral sanctuaries for preserving and applying forces. They can defend in sanctuaries or attack out of them. When defending, the State's forces generally do not employ fixed, contiguous defensive fronts. They may conduct limited-duration operational- and tactical-level offensive actions to prevent buildup of intervening forces, to facilitate the defense, or to take advantage of an opportunity to counterattack.

6-22. When State forces can create a window of opportunity or exploit opportunity created by natural conditions that limit or degrade enemy capabilities, they move out of sanctuary and attack. They try to force the enemy to operate in areas where the State's own long-range fires and strike operations can be most effective. They use windows of opportunity to destroy key enemy systems or cause mass casualties. If these fires and strikes change the balance of forces, State forces can exploit success with decisive offensive maneuver. Otherwise, they go back into sanctuaries.

6-23. The State uses flexible and unpredictable force structures task-organized for particular missions. Forces may be combined arms, joint, interagency, and possibly multinational. The State may fully mobilize all available means to create large conventional force and paramilitary capability in support of adaptive operations. Full mobilization involves all military and paramilitary forces, including militia. During adaptive operations, the State will use conventional forces in adaptive ways. It will also employ unconventional and specialized forces tailored to the needs of combat against an extraregional force with technological overmatch. Operations may also involve various types of affiliated forces.

6-24. In military terms, the extraregional force might have technological overmatch, but may not be able to apply it fully against the State. This can occur when successful strategic operations cause the extraregional power to impose political limitations on its participation in the conflict. Such political considerations can limit the amount of forces being brought to bear, the timetable for deployment, and the application of forces. The State can change the nature of conflict to something for which the extraregional force was not prepared or optimized. The enemy's organization and capabilities may be mismatched to the situation. The State's forces may be able to deplete enemy force not through attrition, but through forcing them into noncombat commitments. The State can also affect enemy capabilities by attacking, threatening, or infiltrating the host-nation contractors on which the extraregional force depends.

6-25. The types of adaptive actions that characterize "adaptive operations" at the strategic level can also serve the State and its Armed Forces well in regional or transition operations—at least at the tactical and operational levels. However, once an extraregional force becomes fully involved in the conflict, the State will conduct adaptive actions more frequently and on a larger scale.

## COMMAND AND CONTROL

6-26. Planning and preparation for adaptive operations begin with the original plan for the strategic campaign, in the form of branches and sequels. During regional and transition operations, the State refines plans and carries out further preparation for adaptive operations.

6-27. Compared to regional or transition operations, adaptive operations rely even more on centralized planning and decentralized execution. Commanders at various levels determine the exact means of execution based on the State's strategic goals, missions received from higher headquarters, assessment of the enemy's capabilities and intentions, and conditions at hand. When overmatched in conventional power, the State expects its commanders to seize opportunity, tailor organizations to the mission, and make creative use of existing capabilities even more than they did in regional and transition operations.

6-28. There are two essential problems the State must solve if it must fight an extraregional opponent. First, since this enemy is a modern force, the State will be at a disadvantage in protecting its own $C^2$ against a more technologically able enemy. Secondly, State planning is driven and developed from the top. Yet, the State must assume there will be disruption to highly centralized command and control. Equally important, adaptive operations also require decentralized execution because the State believes the only way to match or control the tempo of a modern force is to delegate operational and tactical control to the lowest levels of command. Thus, the State's $C^2$ concept still emphasizes the four essential elements: objective, opportunity, method, and end state.

6-29. The State realizes that, in the event of extraregional intervention, it is particularly vulnerable to attacks on its communications system. It attempts to resolve this problem by using commercial digital communications that it can encrypt. The $C^2$ concept, however flexible in theory, is nonetheless a vulnerability. Despite a $C^2$ concept that encourages initiative at all levels, the State does have communications requirements, and it prefers assured communications. It attempts to assure communications by a number of means, including the providing of cellular or digital phone technology to the tactical level. In sanctuary-based adaptive operations, it will pre-position and install fiber-optic communications cable and make other hard-wired arrangements. It understands that there is risk inherent in this approach, but believes that the tempo of operations it is able to sustain will offset the risk.

6-30. The State concept for $C^2$ takes into account the State's vulnerabilities as well as the vulnerabilities it perceives in its potential adversaries. It stresses planning centrally, but encourages flexibility and initiative in decentralized execution. This concept both mitigates State weaknesses and enables it to accelerate its decision making and accelerate or slow the pace of its operations as required. Equally important, the State believes it can operate effectively at a tempo that will challenge potential extraregional opponents. In sanctuary-based adaptive operations, it will afford itself more robust $C^2$ by hard-wiring to the extent possible. It believes this will enable its militia forces and regular forces to act in concert effectively, while accounting for the fact that its militia forces are less able to execute its fundamental approach to command and control.

## INTERNAL SECURITY FORCES

6-31. Internal Security Forces resubordinated to Supreme High Command provide support-zone security, collect information on foreign organizations and spies, and protect the President and other high-ranking State officials. Some of the paramilitary Internal Security Forces have combat potential to conduct defensive operations if required. Internal Security Forces also ensure that mobilized militia forces remain loyal to the State.

## ARMY

6-32. The Army will seek to conduct adaptive operations in circumstances, opportunities, and terrain that optimize its own capabilities and degrade those of the enemy. It will employ a force that is optimized for the terrain or for a specific mission. For example, it will use its antitank capability, tied to obstacles and complex terrain, inside a defensive structure designed to absorb the enemy's momentum and fracture his organizational framework.

6-33. The Army plans and prepares sanctuaries and defenses throughout State territory in anticipation of intervention. It makes maximum use of prepositioned logistics. Then it tries to dictate when and where combat operations occur. Success depends on two critical factors—transition operations that have allowed the Army to occupy defensive positions, and an enemy driven by a timeline for completion.

6-34. Various State forces engage the enemy simultaneously to his operational and even strategic depth. The goal is to present the enemy with a nonlinear, simultaneous battlefield, allowing him no sanctuary. This does not allow the enemy to have any secure rear areas. The State tries to force the enemy to commit forces at very low percentage of combat strength, possibly with no reserve, before he has built up overwhelming forces on the ground. The Army attempts to control enemy movement, using natural and manmade obstacles and exploiting the presence of civilians on the battlefield. Overland routes through complex terrain are obviously vulnerable to attack. Army forces use the enemy's predictable patterns of operation to set up raids and ambushes. They look for opportunities to conduct fast, surgical attacks or counterattacks and then return to sanctuary.

6-35. During dispersed operations, substantial gaps may exist between the positions of units. In these gaps, the State will use precision long-range fires or Special-Purpose Forces (SPF), insurgents, and militia to destroy key systems, cause politically unacceptable casualties, harass the enemy, and maintain contact. Some defensive positions may include complex battle positions or heavily defended spaces that are able to fight in all directions within the limitations of the terrain. These defenses are generally tied to complex terrain. The Army conducts counterattacks at all levels, in the gaps or inside defensive positions, to impose delay, inflict casualties, or preserve critical points in the defense. Sanctuary-based operations require extensive planning, preparation, and pre-positioning of forces, caches of supplies, and critical $C^2$ systems. Urban or other complex terrain provides concealment for these assets. The State's forces also try to deny the enemy use of complex terrain that could protect his forces from fires delivered by State forces from their dispersed locations.

6-36. Fighting in urban areas and other complex terrain is manpower- and resource-intensive. The State may have more of both those types of assets readily available in the region than does the extraregional power. When operating in rugged terrain, the older, simpler systems of the State's Armed Forces may function better and more reliably that the high-technology systems of the extraregional force. In urban areas and other complex terrain, the State can use cheap, plentiful, but lethal infantry systems to destroy enemy platforms. It can also affect the extraregional force's dominance by acquiring niche technology and achieving technological surprise.

6-37. The extraregional enemy prefers to use his technological advantage in the form of long-range, standoff engagement. The State can force such an enemy to engage in close, dismounted combat by conducting the fight in urban or other complex terrain. It can also keep the enemy from using his standoff capability by using the civilian populace and third-party noncombatants as shields or locating the State's forces and systems close to sites that are culturally, politically, economically, or ecologically sensitive, especially in invaded territory.

6-38. The advanced $C^2$ and reconnaissance, intelligence, surveillance, and target acquisition (RISTA) systems of the extraregional force are high-payoff targets for computer warfare, information attack, or physical destruction. Disruption or denial of these resources at critical times can degrade the enemy's situational awareness and real-time intelligence ability. Attacking a critical ground-based $C^2$ or RISTA node can have a very big payoff for low risk and relatively low investment, using common jamming systems and other technologies readily available on the open market. Thus, the State places high priority on identifying and locating enemy $C^2$ and RISTA assets, so that it can destroy, deceive, or manipulate them to its own advantage.

6-39. The Army recognizes the importance of preventing or at least delaying enemy air superiority. Therefore, the Army maintains its own air defense forces in addition to those subordinate to the Air Force. Army air defense includes mobile air defense units and large numbers of shoulder-fired surface-to-air missiles (SAMs).

## NAVY

6-40. The Navy will continue its efforts to limit or disrupt further enemy access into the region by sea. Navy operations become more offensive. However, it may make more aggressive use of its land-based antiship cruise missiles. The State may risk a surge effort by its remaining naval forces in order to produce politically significant casualties. Sinking a major naval combatant, for example, may be worth the risk and effort. The Navy could also use its submarine force to insert SPF for direct action against a high-payoff target outside the region.

## AIR FORCE

6-41. Against an extraregional enemy, the State starts to lose the air superiority it had enjoyed against regional forces. Still, it can employ its relatively limited air assets during early stages of enemy deployment, to control enemy access or to inflict early casualties before the enemy builds up sufficient air and air defense capabilities to dominate the airspace. It can also save its air forces for a surge effort at a critical point in the conflict. However, the State will not delay use of its Air Force assets until such a surge unless it has means, such as underground shelters, to ensure the survivability of its aircraft on the ground. Deep operations by the State's attack helicopter units against extraregional forces are unlikely except against high-value targets that the State estimates to be worth the expenditure of scarce resources.

6-42. The Air Defense Forces, subordinate to the Air Force, focus their efforts on destroying enemy aircraft, while protecting critical defensive positions and key political and economic sites. The State watched with interest the air campaigns in Operations Desert Storm and Allied Force and concluded that it is unlikely to be able to defend all of its airspace in the event of extraregional intervention. Fundamentally, the State accepts that it may not be able to employ a nationally integrated air defense system (IADS) against a modern extraregional force. A vertically integrated system centrally directed at the national level could be a liability from a $C^2$ standpoint. Thus, the State is prepared to use IADS at sector levels, where units may be hard-wired and do not require as large a footprint physically and electronically.

6-43. The State arrays its Air Defense Forces in air defense sectors and develops air defense ambushes along the most likely air avenues of approach. Within sectors, it may be able to challenge the most modern air forces. In choosing to fight within sectors, it accepts risk, in that air defense sectors present seams in the defenses and may be unable to provide mutual support.

6-44. The State also employs passive air defense techniques including dispersal, deception, and camouflage. To the extent possible, it disperses high-value assets. Sector air defense provides three discrete benefits: facilitating passive air defense, reducing the signature of defensive systems, and enabling the State to mass air defense assets from dispersed sites to protect the most critical targets.

6-45. Air Defense Forces have a combination of passive early warning based on observers and radar systems that will serve them well in the early stages of combat operations against even the most modern opponents. In later stages, the State accepts that it must rely heavily on observers and will lose or at least not be able to use both airborne and ground-based early warning systems with optimum effectiveness.

6-46. The State's concept of air defense is not purely defensive in nature. It also focuses on destroying not only enemy aircraft, but also the $C^2$ systems associated with enemy air operations. Thus, it will also attack enemy AEW (AWACS) platforms, and use air defense jammers, GPS jammers, and other electronic warfare methods to disrupt his air capability.

6-47. During adaptive operations in State territory, Air Force SPF provide air base security. They also can conduct raids against enemy air bases and installations within the region. They may also take part in joint SPF operations coordinated by the SPF Command as part of strategic operations.

## STRATEGIC FORCES

6-48. The State considers the Strategic Forces capability, even when delivering conventional munitions, the responsibility of the National Command Authority (NCA). Unable to mount robust air campaigns, the State can use these weapons to mount an equivalent effort. While willing to use long-range missiles and rockets, the State fully recognizes the international implications of doing so, even with conventional warheads.

6-49. Perhaps the most promising means for the State to mitigate or offset the tremendous technological advantage of an extraregional force is to use of weapons of mass destruction (WMD). The State has invested in very capable ballistic missiles and rockets that enable the Strategic Forces to deliver WMD packages. These weapons do not require great accuracy to achieve results. They do not require sophisticated delivery means either. However, the State does not use WMD casually. Although the State understands that WMD confer advantages in first use, it recognizes that they carry great risks for retaliation in kind. Accordingly, it may risk first use only when the payoff appears to outweigh the potential costs.

6-50. During adaptive operations, the State shows greater willingness to use WMD to extend and divert the enemy. It may conduct WMD attacks against third parties in the region on order to draw enemy attention and resources away from the main fight and to affect international opinion. It may use WMD against the extraregional force, particularly in his rear area.

6-51. Escalation from conventional weapons delivery to delivery of WMD is tightly controlled by the NCA. The State understands that the use of WMD (including large high-explosive warheads) delivered by long-range rockets or missiles represents a deliberate escalation or widening of any conflict in which the State is a participant. Consequently, the State is likely to escalate in stages in order to minimize reaction. For example, it might use Strategic Forces units to deliver NBC weapons inside or outside of the region or theater only after using the same units to deliver long-range conventional weapons. It might use WMD operationally or tactically against troops in the field prior to using them strategically. However, once having crossed the threshold of using WMD, the State may use them against any high-payoff target within the range of its delivery systems. Positioning of its Strategic Forces units and supporting $C^2$ nodes could, if detected, serve as indicators of the State's intentions. Therefore, the State will seek to conceal the location of these forces and nodes.

6-52. Strategic Forces that are capable of delivering WMD are the exception to the State's $C^2$ concept of being able to operate in the absence of assured communications. Since the State does not believe that first use of chemicals against units in the field will provoke a nuclear response, it is less rigid than other nations in the control of chemical release. The NCA may pass chemical release authority to operational or even tactical levels. That is not the case in employing NBC warheads on any national- or theater-level systems or systems that have a strategic consequence.

6-53. Nuclear weapons would almost surely provoke response in kind. So the State will probably avoid the use of nuclear weapons against an extraregional power unless survival of the regime or the nation is at risk. Among NBC weapons, the State is more likely to use biological or chemical weapons against even an extraregional enemy, particularly if the enemy does not have the capability to respond in kind. Biological weapons, particularly those of high infection rate at relatively low rates of mortality, offer some advantage, particularly if plausible deniability can be achieved. The State is most likely to use chemical weapons, particularly if it can mitigate the risk of retaliation. During the course of armed conflict, the State may intentionally or unintentionally cause the release of toxic industrial chemicals into the environment. Fuel-air explosives and large conventional warheads offer decided advantage and do not necessarily trigger similar responses from the enemy. The State is particularly likely to use these conventional weapons during adaptive operations when its forces are widely dispersed.

## SPECIAL-PURPOSE FORCES COMMAND

6-54. During adaptive operations, the State may increase the level of SPF actions in the enemy rear area. Together with Strategic Forces, the national-level SPF Command provides the State the ability to attack both regional and extraregional enemies throughout their strategic depth. Strategic reconnaissance by SPF in support of national intelligence requirements is an essential element of access-control operations. In addition to its own direct action against enemy forces and installations, the SPF Command can also support terrorist and irregular forces operations. While SPF belonging to other service components are designed for use at the operational level, the SPF Command's organic forces provide a regional and global strategic capability. The SPF Command has the means to control joint SPF operations involving Army, Naval, Air Force, and/or Internal Security Forces SPF, as required.

6-55. The State has trained SPF as alternate means of delivering nuclear, biological, or chemical munitions packages it may develop for them. This assures redundancy with Strategic Forces' delivery means and provides a worldwide strategic means of WMD delivery. With SPF delivery means, the range of WMD is not limited to the range of the missiles of the Strategic Forces. This is yet another example of a low-end investment that promises a high payoff. However, the use of SPF deployed globally to deliver WMD would require the same NCA-level authorization as delivery of WMD by the Strategic Forces.

## AFFILIATED FORCES

6-56. The State may form alliances with non-state actors, including insurgents, terrorists, or even drug and criminal organizations. These affiliated forces do not necessarily agree with all the State's policies and goals, but may just be against the intervention of an extraregional force whose influence they resent and resist.

6-57. Transnational corporations operating in the State or the region often have their own private security forces that can act independently or as affiliated forces. Sometimes these are highly capable paramilitary forces, possibly including armored vehicles.

6-58. The State may foster insurgency operations against enemy forces in any portion of the region that enemy forces may occupy and in the enemy's operational rear areas. Insurgency operations can divert enemy attention and assets from the main conflict.

6-59. In addition to other types of affiliated forces that may be present in regional or transition operations, adaptive operations may involve partisan operations within parts of the State contested by extraregional forces.

## PARTISAN FORCES

6-60. When the territory of the State is partially overrun by extraregional forces, yet another kind of forces can become involved on the part of the State—partisans. Partisan forces are an important element in the State's concept of total war. Whether or not partisans are considered "affiliated" forces depends on their level of integration into the operations of the regular Armed Forces. Partisan operations typically are conducted by militia units, augmented by civilians and remnants of regular military units. Partisans are "invisible infantry," indistinguishable from the civilian populace.

6-61. If the State has insufficient forces to defend against an invader by "conventional" means, its leadership plans to integrate partisan actions in an attempt to prevent occupation of State territory and to eventually force the invader to withdraw. Partisans task-organize their available forces based on the mission. Many of the tactics, techniques, and procedures used by partisan forces are similar to those used by insurgents.

6-62. The aim of partisan operations is to defeat the enemy through a series of small combat actions aimed at harassing or attriting his forces and disrupting or destroying his $C^2$ elements, LOCs, or logistics support. Partisans attempt to avoid decisive clashes as long as possible, expand the number of lower-level tactical combat actions, and destroy the enemy by conducting raids and ambushes against his vulnerable points, such as logistics and rear areas.

6-63. Partisans must know the terrain and maintain maximum mobility, adapting quickly to new situations and tactics. Partisan forces have several advantages over the enemy. Primary among these are the partisans' extensive knowledge of the State's terrain and their ability to successfully execute their missions in adverse weather conditions. Their familiarity with the terrain allows them to use urban areas and other complex terrain to their advantage. Partisans familiar with the terrain and possessing appropriate equipment can be quite effective in this environment. They are more agile than regular forces.

6-64. Militia units participating in partisan operations may have been bypassed, fragmented, or heavily attrited, or may not have had time to fully mobilize. The effectiveness of these forces and the degree of their integration into regular Army operations varies widely. For example, militia units that have been fully mobilized and previously integrated into regional operations would be more effective as partisans than would previously unmobilized units fighting without knowledge of the missions and plans of regular Army units. Militia units may be augmented by remnants of regular Army forces that have been bypassed, weakened, or attrited. As a rule, partisan units with such augmentation achieve a higher degree of integration into the operations of other regular Army forces by virtue of their familiarity with the missions and plans of a higher headquarters. However, these units are most likely to conduct operations against smaller enemy combat units than do their regular Army counterparts.

6-65. Partisan operations may also include civilians, augmenting militia forces. Enemy forces may have destroyed or bypassed the homes of these civilians. Their degree of integration into the operations of regular Army or militia units is normally low, and they are poorly equipped to engage enemy combat units. However, they can still be effective against enemy logistics and $C^2$ facilities and LOCs.

## STRATEGIC OPERATIONS

6-66. During adaptive operations, the State uses all its instruments of power to deny the extraregional enemy any sanctuary in the region or in his strategic depth. Access-control operations and strategic attack of the enemy LOCs and rear are essential to success. Unable to conduct strategic bombing campaigns, the State may use other means to attack targets in its enemy's homeland or along the LOCs into the theater. The State will coordinate its attacks on enemy infrastructure or even civilian targets with perception management efforts to convey the view that these terrorist-type attacks are no worse than enemy bombing campaigns.

6-67.The Armed Forces conduct operational and tactical lines of operation that accentuate what the State is doing at the strategic level. In most cases, operational and tactical efforts are focused on not being defeated, which offers no decision. Therefore, the State needs its strategic operations to be decisive.

6-68. Concurrent with adaptive operations, the State could also conduct or support insurgency operations either within the region in support of its other operations or outside the region. The purpose of such insurgency operations could be to draw an extraregional enemy's attention and resources from the State's main effort or merely to harass the State's adversaries.

6-69. The State will instigate and support terrorist attacks and/or conduct SPF direct action against the extraregional power, even in that power's homeland. It can also use these types of attack against an extraregional force's coalition partners and allies in order to force them out of the war. Depending on the situation—how the war is going or how the State perceives it is going—the State might use WMD at any time against any conceivable target it can reach. The long-range missiles and rockets of the Strategic Forces are powerful political and psychological tools, whether armed with conventional or WMD munitions. With SPF and terrorist delivery means, the range of WMD is not limited to the range of the missiles of the Strategic Forces.

## Chapter 7

# Force Design, Mobilization, and Sustainment

The State strategic logistics complex is the foundation for the logistics system. Fundamental to the logistics concept are the twin notions of total war and all means necessary. As a consequence, the State fully integrates civilian and military components of both its materiel and service industries. Thus, the State strategic complex includes the national industrial base with its supply points, distribution centers, arsenals, plants, manufacturing facilities, medical support, and personnel support centers. The national industrial base is capable of building everything from small arms to nuclear-capable missiles. However, while the State has the ability to design, produce, and field weapon systems, there are some serious qualitative production and integration shortcomings.

The State logistics system is designed to provide continuous support to the civilian populace while simultaneously supporting military forces from the strategic level to the individual fighting unit. The State's national security strategy requires that the Armed Forces and the entire population be constantly prepared for the sudden outbreak of war or natural disasters. The State continues to make major improvements in all aspects of its logistics system. This includes an increased emphasis on support-zone security and plans for stockpiling war materiel throughout the country.

## STRATEGIC INTEGRATION DEPARTMENT

7-1. In planning and execution, the Minister of National Security establishes priorities and assigns responsibilities for the Ministry of Defense (MOD) as well as the civilian ministries. The Strategic Integration Department (SID) is his executive agent, responsible for integrating all the instruments of national power under one cohesive national security strategy.

7-2. The State has an adequate labor supply to meet its civilian and military manpower requirements. However, the SID must ensure that the military does not employ personnel such as security, medical, technical specialists, truck drivers, machinists, and mechanics to the extent that the economy becomes paralyzed. During peacetime, the State may be unable to fully man the military with critical professional and technical specialists to maintain an elaborate support structure. Thus, the Armed Forces may experience a shortage of doctors, engineers, computer programmers, electronic technicians, and other support professionals. Once the country has been mobilized, however, these critical professionals are detailed into the military structure to augment existing professionals.

7-3. A primary SID objective is to ensure the national industrial base is responsive to both the civilian populace and the military during the strategic campaign. For example, the SID may order stockpiling of critical civilian supplies and materiel in anticipation of economic sanctions imposed by an extraregional power and/or the United Nations. Thus, the impact of the economic sanctions is minimized or has a limited effect.

7-4. The SID, in coordination with the MOD, must determine the functional types and levels of civilian support to the military the State can afford without placing the success of the strategic campaign at an unnecessarily high risk. The scope of the support is limited only by the availability of resources and the ability of the SID to reach a consensus across ministerial lines. The SID designates a single point of contact or liaisons within each ministry to coordinate the activities related to civilian and military support during peacetime, mobilization, and war. This ensures the timely and efficient activation and integration of the civilian populace and the Armed Forces.

7-5. Within the MOD, the SID works closely with the Organization and Mobilization Directorate of the General Staff. That office, one of the most powerful in the government, determines requirements, establishes priorities, and resolves competition for resources in the Armed Forces. The Organization and Mobilization Directorate, in turn, works closely with the Ministry of Finance and Economic Affairs. The objective is to assure that the acquisition of materiel and the development of stockpiles to support sustained operations meet the national priorities established in the State. The Chief of Logistics, who heads the Logistics Directorate of the MOD, executes logistics plans in response to the SID during transition to war and conducts planning with the Organization and Mobilization Directorate.

7-6. The SID also works directly with the Ministry of Finance and Economic Affairs to ensure that economic policies meet a variety of needs. These needs include the financing of the infrastructure and other capital projects or developmental programs. The intent is to avert a major economic or political crisis. Thus, the State's leadership is afforded reasonable economic and political stability at all times.

7-7. The SID coordinates with the Ministry of Public Information to encourage constructive public support for the State's policy objectives and to unmask and counter hostile attempts to distort and frustrate the State's policies and programs. The wide range of information campaigns involves all of the internal media sources.

7-8. The SID develops programs and coordinates humanitarian relief activities to support those who suffer from natural disasters such as earthquake, famine, flood, and drought. The effort is conducted, often in conjunction with those of other nations and nongovernmental organizations, to quickly alleviate the effects of disaster and reduce human suffering.

## FORCE DEVELOPMENT AND INVESTMENT STRATEGY

7-9. The State's concept of force design has two major components: force development and investment strategy. The State develops its forces and employs an investment strategy to deal with perceived threats and opportunities. Its overall approach to force design is the product of a thoughtful analysis of its strategic means in relation to internal threats, regional opponents, and the potential for extraregional intervention. This analysis considers both the advantages the State enjoys and the disadvantages it has in comparison to potential opponents.

7-10. The primary objective is to acquire technology to sustain or increase the economic growth of the State. There are three priorities of technological development. The first priority is the development of technology to support the State's infrastructure. The second priority is the development of dual-use technology for use to support both the civilian and military sectors. The third priority is technology unique to military applications.

## GOALS

7-11. The State is pragmatic in the design of forces and makes few high-cost, high-technology investments beyond those that assure internal security and regional dominance and enable the State to contest access to the region by potential extraregional opponents. Where possible, it also seeks the capability to deny its opponents sanctuary from the tactical to the strategic level. In sum then, the State's goals are the capabilities to preserve its own regime, exert influence in its region, and contest access to the region.

> **Force Design Goals**
>
> The State's force design goals are to allow it to—
>
> - Preserve its own regime.
> - Exert influence in its region.
> - Contest access to the region.

7-12. For dealing with internal threats, the State maintains a variety of internal security forces. These run the gamut from national security and border guard units to national, district, and local police units. Some of these units are paramilitary forces (organized along military lines and in some cases equipped with heavy weapons and armored vehicles. Thus, they can provide combat potential to conduct defensive operations or otherwise supplement regular military forces if required.

7-13. The State invests over the long term to assure strong conventional forces able to overpower opponents in its immediate region, where its Armed Forces enjoy relative strength. However, it does not attempt to match the capabilities of likely opponents from outside the region. Accordingly, it fields a wide range of capabilities, from state-of-the-art systems to obsolescent or even obsolete systems that have been modernized to the extent possible. It places a premium on adaptive and innovative application of existing systems as the only way in which it might be able to match the most modern armies in the world today. It also develops and invests in special-purpose forces (SPF), information warfare, and selected high-payoff modern systems that inhibit outside intervention or have strategic reach.

7-14. Taking the fight to the enemy throughout his strategic depth is part of the State's concept of employing all means necessary. This idea drives the State to fielding SPF with a global reach, as well as Strategic Forces with long-range rockets and missiles.

## FORCE DESIGN PRINCIPLES

7-15. Systems and weapons procurement is driven by the way the State develops forces and, in turn, procurement drives force development. Thus, force design includes three major principles regarding both procurement and force development: whether a system is adaptable, affordable, and deployable.

### Adaptability

7-16. The first principle is whether a system is adaptable—that is, can the system serve multiple purposes, and can it be modernized effectively by upgrades from the supplier or by applying upgrades developed indigenously or procured from a third party? Wherever possible, the Armed Forces also try to field forces and systems mounted on a common chassis. This approach allows relatively low-cost upgrades and direct improvements to force effectiveness.

### Affordability

7-17. The second principle is whether the system is affordable—that is, is the payoff likely to offset the cost? More than any other consideration, this is essential in the State's thinking. For example, ballistic missiles may have a fairly high unit cost, but have high strategic payoff. Missile units, while difficult to maintain, are low-density; so the State can assure relatively high-quality units. On the other hand, state-of-the-art fighter aircraft have high unit costs for relatively low strategic benefit. Moreover, the need to maintain highly trained fighter units continues throughout the life cycle of the system. Accordingly, the State tends to invest more heavily in ballistic missiles rather than state-of-the-art fighter aircraft. The intent of the selective purchase of such high-cost technology is twofold. The State may seek a niche that has strategic value, such as ballistic missiles, or may attempt to achieve surprise by introducing unexpected technology.

### Deployability

7-18. The third principle is whether the system is widely deployable—that is, can the State field it in significant numbers? Another factor driving force design is the State's concept of total war and applying all means necessary. The State can field a large number of conventional units with low-cost, low-overhead systems to enable regional overmatch and a relatively large force of conventional units capable of conducting adaptive operations against an extraregional enemy. The concept of the total war also leads the State to field a large number of reserve and militia formations. While most of these reserve and militia formations have little offensive capability, they are capable of defending the State against outside intervention and are easily a match for likely regional opponents.

## ACQUISITION AND PROCUREMENT

7-19. The Acquisition and Procurement Directorate of the MOD is responsible for the oversight and management of the military acquisition and procurement program. Modernization is a critical and constant requirement for the State. The State seeks to obtain a technological or qualitative edge over any regional opponent through modernization of equipment. The overall objectives of the modernization effort are to maintain a deterrent to attacks from a regional opponent and to win quickly in combat. The State recognizes that new technologies serve to create new capabilities in old systems as hybridization provides rapid, exponential improvements in some systems. Therefore, it prefers initially to purchase equipment upgrades that produce hybrid systems, rather than to create or purchase new systems.

7-20. The State focuses on upgrading weapon systems based primarily upon the perceived regional threat. Depending on the threat, it may place a higher priority on upgrades to a particular service of its Armed Forces. For example, it may initially modernize its Navy and then shift emphasis to its Air Force or Army. However, the modernization priorities generally favor the Army. The modernization effort includes incremental hardware and software improvements, new system development, and the use of commercial off-the-shelf components.

7-21. The rapid growth and proliferation of new technology, combined with the modernization effort, allows the State to achieve equality or even an overmatch of the enemy in niche areas. For example, the Armed Forces may have a computerized fire control system that has limitations based on software in the system. Software upgrades can be purchased from other nations or transnational organizations, transmitted or downloaded electronically, and applied on a matter of minutes. Since the basic operation of the system is unchanged, this improvement is transparent to the user. Thus, it improves system performance with no time-consuming retraining or equipment maintenance downtime.

## ACQUISITION PHASES

7-22. The acquisition process is divided into four phases. The *concept exploration* phase consists of competitive, parallel short-term studies. The focus of these efforts is to define and evaluate the feasibility of alternative concepts and to provide a basis for assessing relative merits of these concepts at the next milestone decision point.

7-23. The *program definition and risk reduction* phase is where one or more concepts, design approaches, and/or parallel technologies are pursued as warranted. Assessments of the advantages and disadvantages of alternative concepts are refined. For example, the State may determine that the procurement of commercial off-the-shelf equipment may be in its best interest in the short term. However, the long-term goal is to reverse engineer, develop, and domestically produce the equipment. This phase is also characterized by the prototyping, demonstration, and consideration of early operational assessments of equipment and weapon systems. This effort ensures that technology, manufacturing, and fielding risks are well in hand prior to the next decision point.

7-24. During the *engineering, manufacturing, and development* phase, the most promising design approach is translated into a stable, interoperable, producible, supportable, and cost-effective design. The equipment is produced and fielded in limited quantities to facilitate the incorporation of design improvements and upgrades.

7-25. The fourth and final phase is *production and operational fielding*. The objective of this phase is to achieve an operational capability that satisfies the mission need or requirement. However, the potential for modification or upgrades to the fielded system is continuously evaluated.

## DOMESTIC

7-26. As a rule, the State gradually updates weapons as the overall economy advances, improving existing equipment. Priorities aside, however, it may choose to make opportunistic purchases for the procurement of key technological advances that can produce a qualitative edge over a regional opponent or mitigate the advantage of an extraregional force. When selecting weapons systems, the State considers training, technical, and support-equipment costs required for maintaining and operating the weapon system, in addition to the cost of the system itself. To reduce the overall operating and maintenance costs, the State may procure systems with the intent of placing a majority of them in long-term storage. A few of the systems (along with simulators and training devices) are fielded to designated training units and facilities to maintain established training standards. Thus, the State attains a benefit by reducing the ownership costs of existing systems, extending their life cycle, and maintaining established training standards.

7-27. Whenever feasible, systems and production facilities are dual-use, serving both civilian and military needs. For example, an aircraft plant may be producing commercial transport as well as combat aircraft. The State may purchase or import a foreign-made avionics package for the transport aircraft with a possible use or application in the combat aircraft. This serves to meet two objectives. First it reduces the unit cost of military production, and secondly it facilitates mobilization.

7-28. The State's basic policy is to export or sell arms to obtain both financial and political advantage. The State attempts to undersell its competitors and usually is able to succeed, since its labor costs are lower than in most industrialized nations. However, the bottom line is that money, as well as politics, governs to whom the State sells. It sells arms for foreign exchange that in turn it can use to pay for the acquisition of more technology and weapons systems it cannot produce.

## FOREIGN

7-29. The cornerstone of the State's foreign acquisition strategy is the import of technology and technical assistance with the purchase of foreign systems and subsystems. This includes the licensing for the manufacture and production of systems and their respective components. The primary factor that determines the State's foreign weapons acquisition is the ability of the State to pay in convertible or hard currency, barter for existing goods or services (including food, fuel, raw materials, or illegal drugs), or possibly attain financial assistance from allies. The State may initially purchase much of the equipment for civilian use, but subsequently find military applications. It may make opportunistic investment in high-payoff foreign systems to create technological niches or to lead into domestic development.

## STATE-OWNED FACILITIES ABROAD

7-30. In the complex world of international business, it is possible for the State's wealthy families (or even the State government) to acquire controlling interest or sole ownership of firms located in other countries. In some instances, State seeks a third-party ownership of these firms in order to disguise both the true ownership and purpose of these firms. Thus, the State has access to factories and other enterprises that use human and natural resources of another country to produce goods and services for the State's use. Additionally, this facilitates the State's ability to generate revenue with which it can purchase other goods and services.

## MOBILIZATION

7-31. The mobilization process is another reflection of the dual principles of the total war and all means necessary. For example, the principle of total war emphasizes universal military training and service. Thus, the State is able to meet the personnel requirement for standing forces, significant reserves, and a militia. Planning for mobilization of reserves and militia is the responsibility of the Organization and Mobilization Directorate of the General Staff.

7-32. The Organization and Mobilization Directorate must also coordinate through the SID with other ministries of the government to assure that their planning will meet wartime needs of the Armed Forces. During wartime, the directorate identifies and recommends priorities of effort in coordination with civilian ministries. It deals particularly with those ministries concerned with transportation and the production or acquisition of food, fuel, and materiel.

7-33. The State's Mobilization Law outlines the duties and responsibilities of State ministries and the civilian populace during a State declared emergency. For example, the liability for military service for men starts at age 18 and lasts until their sixtieth birthday. Once the initial training or service commitment is completed, the personnel are transferred to the Armed Forces reserves. Reserve and militia personnel may receive up to 120 days of training per year. The Mobilization Law also permits limited or selective mobilization of resources appropriate for lesser emergencies. Mobilization measures may be applied to specific organizations, agencies, or units. For natural disasters, it facilitates shifting additional personnel to aid certain districts.

7-34. Priorities for acquisition and sustainment, once developed, are adhered to ruthlessly. Nonetheless, the State's growing economy and the limited means of distribution require patience and time, since the State anticipates outside intervention and the need to stockpile equipment and supplies. This is especially so during peacetime. Despite the State's penchant for preparing for war all of the time, it is unable to meet all of its needs quickly. The need to assign production to stockpiling efforts often means that some units may not enjoy high readiness rates during peacetime. Additionally, the State's approach of stockpiling at numerous sites to offset its distribution problems requires such long lead time that the logistics system imposes limitations on the acquisition system.

7-35. During mobilization, the State's logistics system moves very rapidly and smoothly to support military requirements. Generally, the system functions more effectively following mobilization, at least in part due to the setting aside of civilian high-priority efforts. The SID has the requisite authority to enforce national-level decisions. Thus, the State's civilian populace and Armed Forces logistics are afforded the emphasis and resources that the national strategy requires.

# SUSTAINMENT

7-36. The State's concept of the logistics visualizes war throughout the entire country and region. This includes everything from battle positions to the theater or national support zone.[1] The Armed Forces logistics doctrine is a blend of other nations' systems and domestic concepts. This blend can be attributed to the State's previous experience in war, doctrine adapted from foreign sources, and analysis of the operational environment.

7-37. The largest vulnerabilities of the State's logistics system are maintenance of communications and a transportation network capable of sustaining military operations. Therefore, the State must be opportunistic. For example, it emphasizes the use of civilian and captured vehicles, equipment, and supplies to maintain the tempo of military operations. It also places considerable emphasis on improving the efficiency and security of the logistics system. It has increased the depth of forward service areas and increased the mobility and range of logistics units in support of frontline forces.

7-38. For the Armed Forces, all strategic logistics support is coordinated at the national level through the MOD's Chief of Logistics. The responsibilities of the Chief of Logistics are the same during war and peace. These responsibilities include—

- Procuring of personnel, materiel, and services required by the military.
- Preparing the economy and the people to provide sustained support in case of war.
- Ensuring that an uninterrupted flow of personnel, materiel, and equipment reaches the individual fighting unit at the proper place and time.

7-39. Organizations within the national-level military logistics establishment include materiel support and maintenance units, as well as mobilized civilian resources, to include medical personnel and facilities. Some national-level logistics units may be allocated to subordinate commands to reinforce the units forming their logistics bases, while the remaining units are centralized under the General Staff control.

## SUSTAINMENT TENETS

7-40. Given its disadvantages, the State considers that logistics planning is essential to executing the national security strategy. The State relies on the following three tenets: long-term planning, centralized planning and decentralized execution, and interministerial coordination.

**Sustainment Tenets**

The State relies on the following sustainment tenets:

- Long-term planning.
- Centralized planning and decentralized execution.
- Interministerial coordination.

[1] The *support zone* is that area of the battlespace designed to be free of significant enemy action and to permit the effective logistics and administrative support of forces.

### Long-Term Planning

7-41. Long-term planning ensures that the right materiel is developed in support of the military and assures the State is able to sustain operations. The State plans for sustainment in accordance with mobilization of its forces and the types of strategic-level courses of action it anticipates in a strategic campaign. Conceptually, its logistics capability parallels its force design philosophy. For example, mobile forces designed for regional force-projection operations are generally modern and have the most robust combat logistics capability. Logistics plans are primarily developed to sustain military operations against regional threats. Additionally, the logistics plans include sustainment of operations against an extraregional threat.

### Centralized Planning and Decentralized Execution

7-42. To ensure both priority of effort and efficiency in the sustainment process, the State's logistics operations are characterized by the tenet of centralized planning and decentralized execution. Logistics plans are developed at higher levels and executed by units and organizations at lower levels. Centralized planning requires a focal point for logistics planning and resource allocation at all levels. The individual or unit that is that focal point must be constantly aware of requirements and capabilities. Decentralized execution enhances the flexibility to meet local requirements and the rapid reprioritization of support.

7-43. This tenet is critical for support of reconnaissance fires and strike operations. The Armed Forces use reconnaissance fires to attack specific enemy systems in order to destroy or degrade the combat potential of the enemy force. They employ a strike to destroy an enemy formation after setting the conditions for its destruction. The Armed Forces use a series of caches and short-duration facilities to sustain fire support and maneuver forces during these operations.

7-44. This tenet is also important for supporting the deployment of SPF in the enemy's strategic depth or against his lines of communications (LOCs). Generally, SPF are inserted with the requisite munitions and supplies their missions require. Since SPF units are expected to sustain themselves for the duration if their missions, the Armed Forces rely on the careful planning and stockage of supply caches to sustain these forces.

### Interministerial Coordination

7-45. Finally, strategic logistics plans are coordinated across ministerial lines. This is commonly referred to as interministerial coordination. Interministerial coordination by the SID ensures a balanced responsiveness of the national industrial base to both the civilian populace and the military during the strategic campaign.

## LOGISTICS MISSIONS

7-46. Three terms describe how the Armed Forces provide support during peacetime and wartime. The terms for these logistics missions are primary support, area support, and depot support.

7-47. *Primary support* is a mission given to supply, services, transportation, and maintenance units that normally provide support directly to other units. This allows the primary support unit to respond directly to the supported unit's request for assistance or supplies.

7-48. *Area support* is a mission given to supply, services, transportation, and maintenance units that normally provide support to primary support units and other area support units. Lower-priority units may have to rely on area support, rather than receiving supplies and services directly from the next-higher echelon.

7-49. *Depot support* is a mission given to national-level or strategic units that normally provide support to area support units. Depot support operations include the receipt, storage, and issue of war stocks and domestically produced armaments and materiel, and the overhaul and rebuilding of major end items.

## REGIONAL OPERATIONS

7-50. During regional operations, the State generally possesses the advantage of secure LOCs and the latitude to sustain its forces with little to no disruption. In the best case, the State's economy and industrial base may suffer little or no damage from the war. The State may actually increase production of supplies and materiel in a rush to increase national wartime stockpiles of key military and civilian supplies and material prior to any extraregional intervention that could destroy production facilities or disrupt imports.

7-51. Whether the State suffers damage to its domestic production capability or not, it may also increase production in factories and facilities owned by the State or its ruling elite in foreign countries, based on the perceived possibility of intervention of an extraregional force. This action facilitates the State's ability to produce, transport, and stockpile key military and civilian supplies prior to diplomatic initiatives of an extraregional force and/or the United Nations to close down the State-owned foreign facilities.

## TRANSITION OPERATIONS

7-52. During transition operations, the stockpiling of critical civilian supplies and material (such as food, petroleum products, and clothing) becomes the primary focus of the State. The objective is to enable the State to resist economic sanctions imposed by an extraregional power and/or the United Nations. Thus, the impact of the economic sanctions has a limited effect. This affords the State the ability to sustain military operations to attain strategic objectives and satisfy the basic needs of the populace with minimal impact. The State may gradually begin the rationing of civilian supplies and services.

7-53. The State's militia units may be employed in primarily defensive roles during transition and adaptive operations. They are generally dependent on stockpiled supplies and logistics units of limited mobility. During transition operations, militia units may receive top priority in distribution and transportation of supplies in order to facilitate their execution of missions in support of the strategic campaign plan.

## ADAPTIVE OPERATIONS

7-54. The State primarily shifts its emphasis to an area support mode during adaptive operations, since it no longer possesses the advantage of secure LOCs. For example, it may use all modes of transportation (air, rail, waterway, and road) sparingly due to the Armed Forces' inability to control the State's airspace. Thus, military units and the civilian populace are locally sustained through short-duration facilities such as supply points, caches, and depots. These facilities are capable of accommodating a majority of the demands placed on the distribution system for activities located in their respective support zone.

7-55. Diplomatic initiatives may impact on the State's ability to procure supplies and materiel produced in State-owned factories and facilities based in foreign countries. However, the State remains opportunistic during adaptive operations. If the diplomatic sanctions are not imposed, the State can continue to produce, transport, and stockpile key military and civilian supplies. It could seek to transport supplies and materiel under foreign-flagged commercial transportation as well as through the supply networks of underground or criminally-based activities. The SID normally imposes the rationing of civilian supplies and services. Internal security forces are generally vigilant in their efforts to reduce the impact of black marketing of supplies on the civilian populace.

# MATERIEL SUPPORT

7-56. The Armed Forces materiel support system comprises a mix of very modern and unsophisticated capabilities that vary depending on the priority of the supported units. Generally, high-priority or elite units enjoy the benefits of a robust materiel support system that affords a higher degree of flexibility and responsiveness to rapid changes in plans. For such units, the system may be fully automated to track requirements and control the issue of supplies. Less capable units (including reserve and the militia forces) typically have little or no automation support. Both types of materiel support system are based on allocating supplies and services to units in order to accomplish mission objectives. However, the aim of the State is to continue the upgrade of its less capable units to a robust supply system capable of sustaining the force in all environments.

7-57. *Supply* includes actions to acquire, manage, receive, store, and issue the materiel required to equip and sustain the force from mobilization through deployment, combat operations, and recovery into the State homeland. The allocation of supplies is based on the unit mission, supply reports, and the availability of supplies. During peacetime, the Armed Forces operate under the "pull system" of supply. For example, personnel in the field may request material from a depot where it must be picked up and delivered to the field.

7-58. The State's concept of *services* includes all troops, installations, and duty positions that perform logistics support for combat arms units. Such services are not specific to the ground forces, but support other Armed Forces components as well.

7-59. During wartime, however, the Armed Forces operate under the forward distribution or "push system" principle, in which the higher echelon directly supplies and services the next-lower echelon. Supplies and services are delivered directly to subordinate elements using the organic transportation assets of the higher headquarters. Supplies may be procured or obtained from social groups, consumer cooperatives, government farms, or individual citizens, and by coercion or foraging. Lower-priority units may have to rely on area support or even supply point distribution.

## MAINTENANCE

7-60. Maintenance includes actions taken to keep materiel and equipment in a serviceable condition, to return it to service, or to update and upgrade its capability. Since supplies are limited, the Armed Forces stress preventive maintenance, technical inspections, and proper operating methods to extend the life cycle of equipment. The maintenance system is designed to repair vehicles and equipment as far forward as possible. Repair facilities and units move near the scene of combat rather than waiting for damaged equipment to be evacuated to them. Fixed and mobile repair units extend repair capabilities forward into the battle zone and provide service to the customer unit. During wartime, the types of repair performed at each level depend on the situation. Generally, they are of a lesser degree than in peacetime. The Armed Forces classify three categories of repair: routine, medium, or capital.

7-61. Routine repairs—such as replacements, adjustments, or repair of individual components—require a short time to fix. Generally, maintenance personnel do not disassemble major components as part of routine repair. Medium repairs include the minor overhaul of equipment and the repair of individual components requiring a short time to fix. Capital repairs are conducted at depot level and involve the major overhaul and/or assembly of equipment.

## DEPOT FACILITIES AND OPERATION

7-62. Depots are part of the strategic logistics support structure and hold national-level stockpiles. They occupy fixed peacetime facilities, aboveground and underground structures, plus dispersal sites throughout the country. They manage the distribution of war stocks and domestically produced armaments and materiel, and perform any higher-level repair work that is accomplished in country. Examples of these repairs include aircraft instrumentation, optics, and electronics. The depots manage the distribution of consumables such as fuel, food, and other items from the civilian economy. Rocket and missile units, aviation support units, and air defense maintenance units receive direct support from the nearest depot.

> **Depot Categories**
>
> Depots are generally divided into the following categories:
>
> - Area distribution depots.
> - Ammunition depots.
> - Maintenance depots.
> - Medical depots.

7-63. A single depot may have one or more of the above missions. An area distribution depot (ADD) receives, stores, and distributes items for units operating or assigned within the support zone. Major end items may also be stored in an ADD, but normally are stored in a maintenance depot. The materiel stored within an ADD should accommodate a majority of the demands placed on the distribution system for the units located in the support zone. An ammunition depot receives stores, renovates, issues, and demilitarizes munitions of all types. Maintenance depots overhaul major end items and repairable components and, as necessary, perform limited fabrication and manufacturing. All overhaul items are stored at a maintenance depot until disposition instructions are received from the MOD. Medical depots are discussed in the Medical Logistics section of this chapter.

## ABOVEGROUND STRUCTURES

7-64. Aboveground structures range from factory warehouses to aboveground hardened structures. Hardened structures are reinforced for protection against aerial and ground attack. Earth mounded bunkers are an example of an aboveground hardened structure. The State uses extensive camouflage, concealment, cover, and deception techniques to reduce the detection signature of these structures to enemy reconnaissance platforms. The State can also develop sophisticated decoy sites.

## UNDERGROUND STRUCTURES

7-65. Underground structures include shallow buried and deep underground bunkers and complexes. There are cases where the State uses underground storage facilities to house its command and control ($C^2$) complexes and medical facilities. Underground structures are dispersed throughout the country and consist of intersecting tunnels with multiple exits. Some of these exits may lead to either external combat positions or other subterranean facilities. Large camouflaged doors cover the entrances. The camouflage material matches the surrounding rock so closely that one has to knock on the surface to determine the difference. Normally, a complex may extend over a square kilometer or more. Auxiliary casements in the underground facility may hold fuel, water, food, medical supplies, clothing, or life support equipment.

## SHORT-DURATION FACILITIES

7-66. Short-duration storage facilities play a central role in any strategic campaign that may involve intervention by an extraregional power. For example, prior to conducting adaptive operations, the State plans, develops, and builds short-duration facilities for the pre-positioning of equipment and supplies to sustain deployed forces. The State attempts to anticipate outside intervention and plan accordingly. Logistics items are stockpiled or cached in underground caves and dugout holes, tents, or warehouses, and are dispersed over a wide area. These facilities can be considered a mini-supply depot. They also undergo extensive camouflage and concealment to reduce their detection signature.

# TRANSPORTATION

7-67. The State's military logistics planners base their estimates on the use of all movement resources available. These estimates include tactical combat vehicles as well as civilian transportation assets mobilized to move supplies, equipment, and personnel. For example, during mobilization, civilian trucking and bus companies may be organized as militia truck units to provide transportation of cargo and personnel in the State's homeland. The mobility of logistics units must match that of the supported operation. If the logistics support units fail to achieve this, they may jeopardize the overall success of the State's strategic campaign. Traffic management is the responsibility of the Military Transportation Bureau. The bureau is subordinate to the Materiel Support Department under the MOD Chief of Logistics and is responsible for managing defense transportation requirements using military and civilian resources.

## TRANSPORTATION PRINCIPLES

7-68. The principles of movement apply to all military transportation services and remain constant throughout peace and war. Additionally, they apply regardless of the planning level. During wartime, civilian personnel, transportation assets (including farm animals, vehicles, aircraft, and water vessels), and materiel-handling equipment are mobilized to support the war effort.

### Centralized Planning and Decentralized Execution

7-69. Movement control is centralized at the highest level at which commanders charged with providing total logistics support and monitoring the transportation system and infrastructure can exercise it. This requires a focal point for movement planning and transportation resource allocation at all levels. That focal point, whether an individual or unit, must be constantly aware of requirements and capabilities. Decentralized execution enhances the flexibility to meet local requirements and to rapidly reprioritize support.

### Regulated Movement

7-70. All movement is regulated according to command priorities. Movements are not validated, approved, or initiated if any part of the transportation system cannot meet the requirement. Regulating transportation assets and LOCs is required to prevent congestion, confusion, and conflict of movements. Unregulated use of the transportation system can severely hamper the movement of critical cargo and personnel supporting the strategic campaign. Therefore, traffic in the theater is programmed to provide fluid movement throughout the transportation network. The State's internal security forces support movement control through protection of supply routes of movement and key transportation nodes and centers in the State's homeland or in State-occupied territory in a neighboring country.

7-71. A movement program is a directive that allocates the available transport mode capability to satisfy the movement requirements in accordance with the commander's priorities. The program normally contains detailed information concerning origins, destinations, weights, and cube of cargo, and/or types and number of personnel to be moved.

### Fluid and Flexible Movement

7-72. The transportation system is designed to provide an uninterrupted flow of traffic that adjusts rapidly to changing situations. It is flexible enough to meet the changing priorities of a fluid battlefield and reallocate resources as necessary. Adjustments must be made to meet the variations in wartime intensity. For example, when units are in the offense, the transportation system expands to maintain the tempo of the battle. Conversely, when units are in the defense, the system is contracted, the mode changes, and differing cargo priorities may be necessary. Changes in the operational environment necessitate adjustments to operate in varying conditions and operational and/or tactical situations that may dictate the types of convoys and controls established for movement.

7-73. The availability and use of road and rail networks, airfields, inland waterways, ports, and beaches not only allows the transportation system to meet operational and tactical changes, but also provide redundancy within the overall transportation network. For example, if a portion of a road network is destroyed or rendered unusable, the mode could change to rail or inland waterway.

### Maximum Use of Carrying Capacity

7-74.The principle of making maximum use of carrying capacity involves more than just loading each transportation asset to its optimum carrying capacity. Transport capability that is not used in one day cannot be stored to provide an increase in capability for subsequent days. Similarly, a situation allowing fully-loaded transport to sit idle is just as much a loss of carrying capacity as is a partially-loaded vehicle moving through the system. While allowing for sufficient equipment, maintenance, and personnel rest, planners should keep transportation assets loaded and moving as much as the situation permits.

## TRANSPORTATION MODES

7-75. Transportation operations may include motor vehicles, rail, aircraft, and waterway (coastal and inland) transport vessels. The Armed Forces generally uses motor vehicles to move large quantities of general cargo, petroleum products, and personnel throughout the theater. However, waterway transport vessels may be used to move large quantities of supplies and personnel along coastal or inland waterways to remote areas that are not accessible to motor vehicles.

7-76. As requirements for transportation fluctuate, each mode must be properly used to accomplish the commander's objective. For example, air transport is employed if reaction speed is the priority. Motor transport is considered the most flexible surface mode. It provides door-to-door delivery service and an interface with all other transportation modes.

## PERSONNEL

7-77. The MOD establishes policy, assigns responsibilities, and prescribes procedures for personnel readiness issues as they apply to all members and components (standing forces, reserve, and militia) of the Armed Forces. The Manpower and Readiness Department under the MOD Chief of Logistics is responsible for the administration and management of the military personnel system.

7-78. In consultation with the various service component chiefs, the Chief of the General Staff (CGS) provides advice to the Minister of Defense on manpower and personnel issues impacting the readiness of the Armed Forces and the force structure required to support national security objectives. Additionally, the CGS advises the Minister of Defense on the extent to which the major programs and policies of the Armed Forces in the area of manpower conform with strategic plans.

7-79. In addition to recruiting, organizing, equipping, and training, each of the six service components has the responsibility for providing personnel support to its forces. Major combatant commanders exercise authority over assigned forces. This authority allows the combatant commanders to direct and approve those aspects of personnel support necessary to carry out assigned missions and to standardize personnel policies within the command to the extent that such standardization is necessary to carry out missions assigned to the command.

7-80. The State considers people as one of the most critical assets to the success of any military operation. Thorough planning and efficient personnel management directly influences mission readiness. Therefore, the MOD requires service components to resource personnel requirements in a timely manner to support operational requirements. The State views "personnel support" as all activities associated with assignment of personnel against authorized billets and validated individual augmentation requirements, as well as those administrative activities associated with personnel programs within a command.

### EDUCATION AND TRAINING

7-81. The Military Education Directorate of the MOD oversees universal military training programs designed to maintain the population's knowledge of military operations and tactics. It also coordinates with the State Ministry of Education to ensure military training is integrated within the general education system. The National Military Schools and Colleges Department, subordinate to the Military Education Directorate is responsible for the oversight on all military academic training within the country. The schools include the National Military Academy, a National War College, service schools, and specialized schools.

7-82. There are several levels of schools for the individual training of officers, NCOs, and enlisted personnel. They include basic and advanced branch schools; intermediate and higher-level staff schools; and a National War College. The State also has specialized schools to support various special units and/or military occupations. The specialized schools include airborne, SPF, diving, and information warfare. The State also maintains a foreign exchange program with other countries.

### Officer Training and Competence

7-83. The State believes that politically-aware commanders thoroughly grounded in doctrine, organizations, capability, and strategy are essential to the strategic and operational means of executing the State's military art. All military officers adhere to the rule of selfless service to the State. Officer candidates undergo a rigorous academic, physical, and psychological screening process prior to selection for attendance at the National Military Academy. The literacy rate of officers approaches 100 percent. As a rule, the officer corps is comfortable with electronic and automotive technology. The very best officers attend foreign military schools.

### NCO Training and Competence

7-84. The State has established leadership schools for its NCO corps. Most of these schools are oriented toward the development of the leadership qualities of new or junior-ranking NCOs. However, NCOs attending these schools do not necessarily get credit for promotion. The NCOs have a literacy rate of over 80 percent and have some familiarity with electronic and automotive technology.

### Enlisted Training and Competence

7-85. The Combat Training Directorate of the MOD plans, coordinates and supervises combat training in all service components. The State dedicates special schools or units to receive and train newly enlisted or conscripted personnel. Normally, basic training for an enlisted soldier lasts 6 to 10 weeks. Higher-caliber soldiers are identified to receive some type of specialized training ranging from 8 to 12 weeks and sometimes longer. However, most soldiers receive training lasting 4 to 8 weeks as specialists in one field or another. There is a minimum 2-year service commitment for regular force personnel and an 18-month service commitment for conscripts.

7-86. During wartime, soldiers attend 1 to 6 weeks of basic training and then are sent to regular units and integrated into squads or crews. Junior-ranking personnel equipped with training schedules or formal standards established by a higher headquarters train newly accessed soldiers at unit level. Enlisted soldiers have a literacy rate of over 80 percent and possess some familiarity with electronic and automotive technology.

### Unit Training

7-87. The military leadership clearly understands that the training methods of the past are hopelessly inadequate to prepare units for modern warfare. Regional conflicts are expected to be very lethal and may include high-technology weapons. Lessons learned by observing the fate of Iraqi troops in the Gulf War reinforced the need to improve training of the State's Armed Forces. The speed at which the U.S.-led coalition forces cut through the infantry-heavy Iraqi units served to warn the State's leadership of its own possible fate in a future war. Except for a few elite, high-readiness units, most combat units prior to the Gulf War adhered to a numbingly monotonous field-training regimen that expended few resources, took few risks, and produced predictable results.

7-88. The State now has training centers that employ "aggressor" units trained and organized to simulate potential enemy forces in the offensive and defensive combat against the State's Armed Forces. The "aggressors" use electronic countermeasures, mobility and countermobility methods, reconnaissance, counterreconnaissance, and heavy and light forces units.

7-89. Military operations provide the State's military leaders the opportunity to develop and implement new doctrinal and tactical techniques, as well as gain combat experience. They are capable of task-organizing units based on mission, terrain, enemy disposition, and availability of equipment. The task-organized units function effectively, demonstrating a high degree of training in $C^2$ during combined arms operations.

## MEDICAL SUPPORT

7-90. The basic principle of combat medical support is multistage evacuation with minimum treatment by medical personnel at each unit level. They treat the lightly wounded who can return to combat and those casualties who would not survive further evacuation without immediate medical attention.

7-91. The State divides the range of medical treatment into three categories. The first category of procedures includes only mandatory lifesaving measures. The second category includes procedures to prevent severe complications of wounds or injuries. The final category of treatment includes procedures accomplished only when there is a low casualty load and reduced enemy activity.

7-92. In anticipation of an overtaxed combat medical support system, the State's military doctrine emphasizes the importance of self-help and mutual aid among individual soldiers. This concept extends beyond the battlefield to casualty collection points and unit aid stations. Self-help and mutual aid reduces the demands made on dedicated medical personnel, particularly when there is a sudden and massive influx of casualties. Each soldier is required to attend a first-aid training session.

### MILITARY HEALTH SERVICES AND PUBLIC HEALTH

7-93. The Military Health Services Department, subordinate to the MOD Chief of Logistics, is responsible for providing health care to Armed Forces personnel and their dependents. Health care is provided at civilian as well as military health care facilities. The military health services are capable of supporting combat forces in prolonged or major military operations.

7-94. The military health services also benefit from health services available through the civilian sector. The Ministry of Health is responsible for managing public health services and regulating private medical services. The ministry's goal is to ensure that all citizens receive health care. Preventive medicine, including staff training, program management, vaccinations, public education, and improved water supply, is a priority. The ministry seeks to extend primary health care centers, ensure availability of pharmaceuticals, and control population growth. Primary, secondary, and tertiary medical services are offered through a network of facilities. Secondary and tertiary care services are also available through university hospitals under the Higher Education and Scientific Research Directorate of the Ministry of Education.

7-95. During peacetime, members of the reserve and militia receive their medical care through the government-sponsored National Social Health and Security Fund. This fund, financed by employee, employer, and government contributions, provide medical insurance to the civilian populace. It seeks to control health care costs and gradually reduce the numbers of patients going abroad for health care by improving services within the country. The fund offers primary health care through numerous clinics and hospitals. During wartime, the reserve and militia receive medical support through the military health care system.

## INTERNATIONAL MEDICAL ASSISTANCE

7-96. Nongovernmental organizations (NGOs) and private volunteer organizations (PVOs) often operate within the borders of the State to provide humanitarian assistance. However, there are cases where the assistance is somewhat restricted by security problems. International aid consists primarily of materiel, funds, and personnel (local and foreign).

## MEDICAL FACILITIES

7-97. The best medical facility is the Central Army Hospital. During peacetime, military personnel receive treatment at this hospital, which also is designated as one of the emergency medical care facilities for foreign diplomats and their families. During wartime conditions, military personnel are treated at all of the major civilian hospitals. The State has designated some of its major university hospitals as emergency medical care facilities for foreign diplomats, their families, and tourists. This action ensures consistent high-quality medical staffing, care and treatment. These facilities are also staffed with foreign medical personnel. A majority of medical facilities or clinics in the outlying areas have sufficient numbers of trained personnel, supplies, and reliable electric power and water. The facilities also contain high-quality, sophisticated, domestic and imported medical equipment. The pharmacies are stocked with high-quality domestic, and foreign-produced pharmaceuticals.

## MEDICAL LOGISTICS

7-98. The medical logistics system operates on a "pull system." Personnel in the field request medical material (including repair parts for medical equipment) from a medical depot where it must be picked up and delivered to the field. Normally, medical supplies are transported forward on cargo carrying transport vehicles, water vessels, or aircraft. However, ground ambulances returning to forward areas may assist in transporting medical supplies forward. A medical equipment maintenance unit at the medical depot provides all medical equipment maintenance.

## QUALITY OF MEDICAL PERSONNEL

7-99. The skills of the State's medical personnel are comparable to those of their foreign counterparts. The State's military doctors are joined in wartime by the mobilization of the bulk of the country's physicians. Many medical personnel have experience treating trauma patients and handling mass-casualty situations. Military physicians are well trained, and the qualifications of most other medical personnel range from good to excellent. All medical personnel receive annual training in advanced trauma life support. After active duty, most of the physicians remain part of the medical reserve and are subject to annual recalls. Once the country has been mobilized, civilian medical personnel are detailed into the military structure to augment existing professionals.

7-100. The training of medical officers, dentists, and pharmacists is conducted at both domestic and foreign specialized medical institutions. Medical personnel also undergo advanced and continuing education through training at the Central Army Hospital. The National Military Academy also has a medical department that is the center of scientific research into theoretical and practical military medical problems. The medical department serves as a major training center for instructors in medical and biological sciences. The medical department coordinates various medical research projects conducted in both domestic and foreign civilian medical institutions.

## CASUALTY HANDLING

7-101. The State has shown success in handling combat casualties. This success stems from emphasis placed on trauma training and close coordination with the civilian medical sector. Evacuation is based on a higher-to-lower method. The next-higher echelon provides transportation for casualties. Each level has specific responsibilities for the care of the sick and wounded. Besides treating the wounded, medical personnel handle virtually all of their own administration, especially at lower levels. As casualties move through the combat evacuation system, medical personnel at each level make effective use of medical facilities by repeated sorting of the wounded (triage). Helicopters are used for all military and civilian search and rescue missions, medical evacuations, and domestic disaster relief flights. During wartime situations, most casualties arrive at a hospital within 6 to 12 hours after being wounded. The evacuation time is reduced to 2 hours during peacetime.

7-102. Treating nuclear, biological, and chemical (NBC) casualties is a standard State trauma protocol. The NBC medical plan is based on three assumptions: mass casualties will occur, casualties will be similar to those that medical personnel have been trained to treat, and medical personnel will be able to treat the casualties in a decontaminated environment. The Central Army Hospital can be converted into a chemical decontamination center within 2 to 6 hours. Most of the remaining major hospitals require 30 days to convert to a decontamination center.

## BLOOD MANAGEMENT

7-103. The Ministry of Health maintains a Blood Management Office to oversee the collection, processing, storage, and distribution of blood (to include liquid blood and blood components) to alleviate the effects of a natural disaster or war. Blood management services are provided to support both civilian and military establishments. The main source of blood to support wartime casualty requirements comes from the collection and processing of blood from the civilian populace during peacetime. The blood is generally stored in prepositioned underground structures throughout the country.

## CIVIL DEFENSE, ENGINERING, AND CONSTRUCTION

7-104. The Civil Defense Directorate is subordinate to the Ministry of the Interior. The directorate determines the assets and the services required to perform peacetime and wartime civil engineering and construction efforts as directed by the SID. During wartime, the work of the Civil Defense Directorate is to a large extent identifying and recommending priorities of effort in repairing battle-damaged facilities and structures (including bridges, storage depots, and airfields) and to minimize or reduce the impact of enemy strikes on the overall infrastructure. This also ensures sustainability of the force to meet the overall strategic goals or objectives.

7-105. Civil defense and engineering programs, directed by the SID, provide for the construction, operation, and maintenance of roads and facilities, and logistics enhancements required to sustain military operations. For example, construction programs may include the building of shelters, warehouses, aboveground and underground structures, road networks, terminals, and hospitals. Engineers and support personnel may operate electric power, sewage treatment, and water and fuel storage and distribution facilities. These personnel may also conduct environmental support operations, provide fire protection, conduct explosive ordnance disposal, provide water purification and disposal, conduct engineer reconnaissance, and provide force-protection construction support.

## LEGAL AFFAIRS

7-106. The Legal Affairs Department under the MOD Chief of Logistics provides professional legal services to the MOD and its staff. These legal services include matters that affect the morale, order, and discipline of the military. The department's responsibilities are to—

- Supervise the administration of military justice within the country and occupied territory.
- Provide legal services to commanders, staffs, and other authorized personnel on all matters involving military law, domestic law, and international law.
- Consult and coordinate with other governmental agencies on legal matters. In particular, the department works in close coordination with the Legal Affairs Directorate in the Ministry of Public Information.

# Glossary

The glossary lists acronyms and terms with Army or joint definitions, and other selected terms. Terms with specific OPFOR-related definitions for which FM 7-100 is the proponent manual (the authority) are marked with an asterisk (*) and followed by the number of the paragraph (¶) or page where they are defined. For other terms, refer to the manual listed.

| | |
|---|---|
| ***access limitation** | ¶3-75 |
| ***adaptive operations** | ¶3-21, p. 6-1 |
| **ADD** | area distribution depot |
| ***administrative force structure** | ¶2-99 |
| **AEW** | airborne early warning |
| **APOD** | aerial port of debarkation |
| **APOE** | aerial port of embarkation |
| ***armed conflict** | ¶3-13 |
| **AWACS** | airborne warning and control system |
| **BTG** | brigade tactical group |
| **$C^2$** | command and control |
| **CGS** | Chief of the General Staff |
| **COE** | contemporary operational environment |
| ***complex terrain** | ¶1-40 |
| ***contemporary operational environment** | p. iv, p. viii |
| ***contemporary OPFOR** | p. xi |
| **CTC** | combat training center |
| **DCSINT** | Deputy Chief of Staff for Intelligence |
| **DOD** | Department of Defense |
| **DTG** | division tactical group |
| ***enemy** | p. ix |
| **EW** | electronic warfare |
| **FG** | field group |
| **FM** | field manual |
| **GNP** | gross national product |

| | | |
|---|---|---|
| **GPS** | global positioning system | |
| **HUMINT** | human intelligence | |
| **IA** | information attack | |
| **IADS** | integrated air defense system | |
| **IRBM** | intermediate-range ballistic missile | |
| **IW** | information warfare | |
| **JP** | joint publication | |
| **LOC** | line of communications | |
| ***military strategic campaign plan** | | ¶3-37 |
| **MOD** | Ministry of Defense | |
| ***national power** | | ¶2-1 |
| ***national security strategy** | | ¶3-1 |
| ***national strategic campaign** | | ¶3-30 |
| ***national strategic campaign plan** | | ¶3-33 |
| **NBC** | nuclear, biological, and chemical | |
| **NCA** | national command authority | |
| **NCO** | noncommissioned officer | |
| **NGO** | nongovernmental organization | |
| **OE** | operational environment | |
| **ODCSINT** | Office of the Deputy Chief of Staff for Intelligence | |
| **operational environment** | | p. iv, JP 1-02 |
| ***operational exclusion** | | ¶3-73 |
| **OPFOR** | opposing force | |
| ***opposing force (OPFOR)** | | pp. x, xi |
| **OSC** | operational-strategic command | |
| ***personnel support** | | ¶7-80 |
| **PME** | peacetime military engagement | |
| **PVO** | private volunteer organization | |
| ***regional operations** | | ¶3-21, p. 4-1 |
| **RISTA** | reconnaissance, intelligence, surveillance, and target acquisition | |
| **ROE** | rules of engagement | |
| **SAM** | surface-to-air missile | |
| **SATCOM** | satellite communication(s) | |
| **SCP** | strategic campaign plan | |

| | | |
|---|---|---|
| ***services** | | ¶7-58 |
| **SHC** | Supreme High Command | |
| **SID** | Strategic Integration Department | |
| **SPF** | Special-Purpose Forces | |
| **SPOD** | sea port of debarkation | |
| **SPOE** | sea port of embarkation | |
| **SSM** | surface-to-surface missile | |
| ***strategic operations** | | ¶3-21, 3-47, 3-103 |
| ***strategic preclusion** | | ¶3-71 |
| ***supply** | | ¶7-57 |
| ***support zone** | | ¶7-36n |
| ***system** | | ¶3-6 |
| ***systems warfare** | | ¶3-7 |
| **TBM** | theater ballistic missile | |
| ***theater** | | ¶3-41 |
| ***threat** | | p. x |
| ***total war** | | ¶3-13 |
| **TRADOC** | U.S. Army Training and Doctrine Command | |
| ***transition operations** | | ¶3-21, p. 5-1 |
| **UAV** | unmanned aerial vehicle | |
| **VIP** | very important person | |
| ***war** | | ¶3-13 |
| **WMD** | weapons of mass destruction | |

# Bibliography

The bibliography lists field manuals by new number followed by old number.

## DOCUMENTS NEEDED

These documents must be available to the intended users of this publication.

JP 1-02. *Department of Defense Dictionary of Military and Associated Terms.* Available online: http://www.dtic.mil/doctrine/jel/doddict/

FM 1-02 (101-5-1). *Operational Terms and Graphics.* 30 September 1997.

## READINGS RECOMMENDED

These sources contain relevant supplemental information.

### ARMY PUBLICATIONS

Most Army doctrinal publications are available online: http://www.adtdl.army.mil

FM 7-100.1. *Opposing Force Operations.* TBP.

FM 7-100.2. *Opposing Force Tactics.* TBP.

FM 7-100.3. *Opposing Force: Paramilitary and Nonmilitary Organizations and Tactics.* TPB.

FM 7-100.4. *Opposing Force Small Unit Tactics.* TBP.

FM 7-100.5. *Opposing Force Organization Guide.* TBP.

FM 7-100.6. *Opposing Force: Worldwide Equipment Guide.* TBP.

# Index

Entries are by paragraph number unless page (p.) or pages (pp.) is specified. After a page reference, the subsequent use of paragraph reference is indicated by the paragraph symbol (¶).

### B

## D

## N

Made in the USA
Monee, IL
15 June 2026

54207762R00089